Science and the Reenchantment of the Cosmos

The Rise of the Integral Vision of Reality

Ervin Laszlo

Inner Traditions
Rochester, Vermont

Inner Traditions
One Park Street
Rochester, Vermont 05767
www.InnerTraditions.com

Library of Congress Cataloging-in-Publication Data

Laszlo, Ervin, 1932-
Science and the reenchantment of the cosmos : the rise of the integral vision of reality / Ervin Laszlo.
p. cm.
Includes bibliographical references and index.
ISBN 1-59477-102-2
1. Parapsychology and science. 2. Cosmology. 3. Science—Philosophy. I. Title.
BF1045.S33L37 2006
501—dc22

2005031698

Printed and bound in Canada by Transcontinental Printing Inc.

10 9 8 7 6 5 4 3 2 1

Text design and layout by Jonathan Desautels
This book was typeset in Sabon, with Adobe Caslon Pro used as a display typeface

To send correspondence to the author of this book send a first class letter to the author c/o Inner Traditions • Bear & Company, One Park Street, Rochester, VT 05767, and we will forward the communication to the author.

Contents

Introduction

At the cutting edge of contemporary science a remarkable insight is surfacing: the universe, with all things in it, is a quasi-living, coherent whole. All things in it are connected. All that happens in one place also happens in other places; all that happened at one time happens at all other times. And the traces of all things that ever happened endure; nothing is entirely evanescent, here today and vanished tomorrow.

The universe is not a world of separate things and events, of external spectators and an impersonal spectacle. It is an integral whole. Unlike the disenchanted world of classical physics, it is not fragmented into material things and the disjointed domains of life and mind. Matter—the kind of "stuff" that makes up particles joined in atoms joined in molecules joined in cells joined in organisms—is not a separate kind of thing, and it doesn't even have a reality of its own. Although it appears solid, in the last count matter is energy bound in quantized wave-packets and these packets are further bound together to create the vast and harmonious architecture that makes up the world. The widespread idea—that all there is in the universe is matter, and that all matter was created in the Big Bang and will disappear in a Big Crunch—is a colossal mistake. And the belief that when we know how matter behaves we know everything—a belief shared by classical physics and Marxist theory—is but sophistry. Such views have been definitively superseded. This universe is more amazing than classical scientists, engineers, and Marxists held possible. And the connectedness and oneness of the universe is deeper and more thorough than even writers of science fiction could envisage.

The current finding of the universe's wholeness is the fruit of sustained investigation, based on observation and tested by experiment. It provides an entirely different image of the world than the mechanistic, materialistic and fragmented image we were taught in school. A cosmos that is connected, coherent, and whole recalls an ancient notion that was present in the tradition of every civilization; it is an enchanted cosmos.

The reenchantment of the cosmos as a coherent, integral whole comes from the latest discoveries in the natural sciences, but the basic concept itself is not new; indeed, it is as old as civilization. In ages past the connectedness and wholeness of the world was known to medicine men, priests, and shamans, to seers and sages, and to all people who had the courage to look beyond their nose and stay open to what they saw. Theirs, however, was the kind of insight that comes from mystical, religious, or aesthetic experience and was private and unverifiable—even if it appeared certain beyond doubt. Now, in the first decade of the twenty-first century, innovative scientists at the frontiers of science are rediscovering the integral nature of reality. They lift the private experiences that speak to it from the domain of unverifiable intuition into the realm of interpersonally verifiable public knowledge.

The emerging vision of reality is more than theory, and it is of interest to more than scientists. It gets us closer than ever before to rending apart the veils of sensory perception and apprehending the true nature of the world. Even in regard to our life and well-being, this is a happy rediscovery: it validates something we have always suspected but in modern times could not express (nor, unless we were poets or lovers, did we even try). This something is a sense of belonging, of oneness. We are part of each other and of nature; we are not strangers in the universe. We are a coherent part of a coherent world; no more and no less so than a particle, a star, and a galaxy. Only, we are a *conscious* part of the world, a being through which the cosmos comes to know itself. This insight is a sound basis for recovering a deeper sense of meaning in life—and a new and more reliable orientation at this crucial juncture in history.

We can live up to our potentials as conscious beings: we can come to know the reenchanted cosmos. Not only is this not impossible, it is not even particularly difficult. Beyond the complex deductions and abstruse mathematics of the new sciences, the basic concept of a coherent, connected, and integral universe is simple and meaningful; indeed, it is beautiful. When we make it our own, we come back to where our fathers and forefathers have been before us. But we come back with more assurance than they have had. Cutting-edge science tells us that we do not delude ourselves: we inhabit an integral, whole universe, and we are part of it. We are at home in the cosmos.

Part One

THE REENCHANTMENT OF THE COSMOS

CHAPTER ONE

The Amazing Coherence of (Nearly) Everything

Whether we like it or not, science enjoys credibility in the eyes of the general public. Most people believe that what science tells us about the world is true, or at least it is likely to be true. Thus, understanding what science tells us about the world is important: it is important both as a possible source of truth about the world and as what most people accept as the truth, or the likely truth, about it.

However, what most people do not realize is that what science says about the world is subject to change, even fundamental change. Indeed, in our day science's concept of the world is changing; it is changing fundamentally. It is not what scientists discussed but a few decades ago, and is far different from what we learned in school. Ours is not the kind of world our common sense would expect: a simple, orderly world where things behave as solid material objects should behave and are either here or there and not in many places at once. Nor is the effect of one thing necessarily limited to just one or a few other things. True, such conditions hold in our immediate surroundings, but they turn out to be limited to certain orders of size and magnitude, and certain dimensions of speed and distance. Beyond these dimensions the world becomes more and more strange. It is with good reason that a widely discussed film asks "what the bleep do we know?" and suggests that it is our consciousness that creates reality . . .

However, even if the way the world behaves is surprising, it is neither arbitrary nor incomprehensible. Scientists have gone a long way toward working out its underlying dynamics and showing that it is not meaningless and haphazard, but has a logic of its own. Notwithstanding the challenge to common sense, the universe proves to be meaningful; indeed, more meaningful than the disenchanted world where inert matter moves impersonally against a background of passive space. The reenchanted world is a harmonious structure where all things interact with all other things and together create a coherent whole. This whole is not a mechanical aggregate, for it is not readily decomposable to its parts. It is an integral whole, where to some extent and in some way all things interact with all other things. The scope of this interaction transcends the hitherto known limits of time and space. It is nearly instantaneous over any known distance and is conserved over any known span of time.

The findings that delineate the new picture of the world come from almost all of the empirical sciences: from physics and cosmology, from the life sciences, even from consciousness research. Although their subject matters differ, their findings have a common thrust. They disclose *interaction* that creates *interconnection* and produces instant and multifaceted *coherence*. The hallmark of a system of such coherence is that its parts are correlated in such a way that what happens to one part also happens to the other parts—hence it happens to the system as a whole. The system responds to the rest of the world as a whole, maintains itself as a whole, and changes and evolves as a whole. It *is* a whole: an integral whole.

THE AMAZING COHERENCE OF THE BODY

An amazing form and level of coherence characterizes nearly everything in the universe, from the largest structures of the cosmos to the smallest particles of the microworld. It characterizes the human body as well.

On first sight, the coherence of a living body stands to reason. Coherence, after all, is a precondition of life itself. If an organism is not to succumb to the constraints of the physical world, all its parts and organs must be precisely yet flexibly connected with each other. Without a high level of coherence throughout this network of connection, physical processes would soon break down the organization of the living state, bringing it closer to the inert state of thermal and chemical equilibrium in which life is impossible. Near-equilibrium systems are largely inert, incapable of sustaining processes such as metabolism and reproduction, essential to the living state. An organism is in thermodynamic equilibrium only when it is dead. As long as it is living it is in a state of *dynamic* equilibrium in which it stores energy and information and has them available to drive and direct its vital functions.

Dynamic equilibrium calls for long-range, quasi-instantaneous connections throughout the organism. The extent of these connections, and the speed with which they spread in the organism, are far greater than scientists had hitherto suspected. The human body, for example, consists of some 1,000,000 billion cells, far more than the number of stars in the Milky Way galaxy. Of this cell population 600 billion are dying and the same number are regenerating every day—over 10 million cells per second. The average skin cell lives only for about two weeks; bone cells are renewed every three months. Every 90 seconds millions of antibodies are synthesized, each from about 1,200 amino acids, and every hour 200 million erythrocytes are regenerated. According to radio isotope analyses at Oak Ridge Laboratories, in the span of a year 98 percent of the atoms that make up the organism are replaced as well. No substance in the body is constant, though heart and brain cells endure longer than most. Yet the substances that coexist at a given time produce thousands of biochemical reactions in the body each and every second, and they are all precisely and almost instantly coordinated so that they maintain the dynamic order of the whole organism.

The vital functions of the body are governed by constant, quasi-instant, and multidimensional correlations. Simple collisions among neighboring molecules—mere billiard-ball push-impact relations—do not suffice. They are complemented by a network that correlates all parts of the system, even those that are distant from one another. Rare molecules, for example, are seldom next to each other, yet they find each other throughout the organism. This is important, for the organism needs to react to stresses and strains as a whole, mobilizing all its resources wherever they are located. There would not be time for an integrated response to occur by a random process of jiggling and mixing; the molecules need to locate and respond to each other specifically, whether they are proximal or distant.

In the living organism order is dynamic and fluid, the body's myriad activities are self-motivated, self-organizing, and spontaneous. The body's coherence extends to every level, from the tens of thousands of genes, hundreds of thousands of proteins and other macromolecules that make up a cell, to the many kinds of cells that constitute tissues and organs. Adjustments, responses, and changes required for the maintenance of the organism propagate in all directions at the same time. All components are in constant and nearly instant communication.

The body's high level of internal coherence makes possible a high level of sensitivity to the external world. In the insect world a few pheromones in the air are sufficient to attract males to prospective mates many miles away. In a human being the eye can detect single photons falling on the retina, and the ear can detect the motion of single air molecules. The mammalian body responds to extremely low frequency electromagnetic radiation, and to magnetic fields so weak that only the most sophisticated instruments can register them. Such sensitivity is only possible when a large number of molecules are coherently linked among themselves.

Biophysicist Mae-Wan Ho remarked that no matter how diverse they are, the parts of the organism act like a good jazz band, where every player responds immediately and spontaneously to the improvisations of all the others. The "music" of the body ranges over more than seventy octaves. It consists of the vibration of localized chemical bonds, the turning of molecular wheels, the beating of micro-cilia, the propagation of fluxes of electrons and protons, and the flowing of metabolites and ionic currents within and among cells through ten orders of spatial magnitude. As long as the organism is alive, this music never ceases. It expresses the harmonies and melodies of the individual organism with a recurring rhythm and beat, and with endless variation. The organic jazz band can change key, change tempo, and even change tune, as the situation demands, spontaneously and without delay. There is a basic structure, but the real art is in the improvisation, where each and every player enjoys maximum freedom of expression while remaining perfectly in tune with all the others.

THE AMAZING COHERENCE OF THE UNIVERSE

The coherence of the body is staggering, yet even this level of coherence is not unexpected in a living system. Life could not exist unless the organisms that manifest it are through-and-through correlated and coherent. But until recently we had no inkling that physical nature—the universe as a whole—could be similarly coherent. Yet this is the case. The cosmos, it turns out, is not a collection of bits of matter, clumped randomly into galaxies, stars, and planets. It is an integral system in its own right.

According to the so-called standard model of cosmology, the universe we live in originated in a cosmic explosion about 13.7 billion years ago. That explosion, known as the Big Bang, occurred in a fluctuating sea of virtual energies: the quantum vacuum. A region of this vacuum became unstable, creating a fireball of unimaginable heat and density. In the first milliseconds it synthesized all the matter that now populates cosmic space. The particle-antiparticle pairs that emerged from the vacuum collided with and annihilated each other, and the one-billionth of the originally created particles that survived (the tiny excess of "matter" over "antimatter") now makes up the physical furnishings of the universe.

After about 200,000 years the particles decoupled from the radiation field of the primordial fireball, space became transparent, and clumps of particles—mostly in the form of hydrogen atoms—established themselves as distinct elements in the cosmos. Due to gravitational attraction they condensed into gigantic swirls that solidified as galaxies. The atoms that made up the physical substratum of these swirls condensed further: the first stars appeared about 200 million years after the Big Bang. Some stars spun off planets, and on the planet we inhabit, the first signs of life appeared about 3.7 billion years ago.

This scenario of cosmic evolution is well established, but the full scope of the coherence now reigning in the universe has only recently come to light. This coherence has surprising aspects and unexpected dimensions. First of all, the "universal constants" (the force of gravitation, of electromagnetism, of nuclear fields, as well as the size of the electron, the neutron, and other fundamental particles) are so finely tuned to each other that the universe could produce active stars that shine long enough to create physical and chemical conditions on some of their planets that are just right for the evolution of life. Even the basic parameters of the universe are amazingly coherent. The mass of elementary particles, the number of particles, and the forces that exist between them are all adjusted to favor specific harmonic ratios (such as the ratio 10^{40}) that recur over and over again.

And this is still not all. It turns out that galaxies evolve nearly uniformly in all directions from Earth, a finding that is surprising, because not all parts of the universe are connected by light, the fastest form of physical signal-transmission according to Einstein's theory of relativity. At its periphery the universe expands beyond the 13.7 billion light years that rays of light could have traveled in the time that has elapsed since the Big Bang—and yet galactic evolution is consistent throughout the universe.

The "flatness" of space-time points to another form of coherence in the universe. In the absence of the massive particles that exert gravitational attraction, space-time turns out to be exactly Euclidean or "flat" (the kind of space where the shortest distance between two points is a straight line). This indicates that the Big Bang that had set the basic parameters of the universe must have been stupendously finely adjusted—because if it had produced just one-billionth more matter than it did, space-time would be positively curved like a balloon, and if it had produced one billionth less, it would be negatively curved like a saddle.

The coherence of the cosmos is intrinsically related to the nature of its fundamental particles, and these turn out to be strange beasts indeed. Although in some experiments particles appear to be discrete corpuscular packets of energy, they also turn out to be continuous with each other and remain thoroughly "entangled"—that is, interconnected—over all finite times and distances. The particles have both corpuscular and wave aspects, and until they are observed or measured, have no definite characteristics but exist simultaneously in several states at the same time. These states are not "real" but "potential"—they are the states the particles can assume when they are observed or measured. (It is as if the observer, or the measuring instrument, fishes the particles out of a sea of possibilities. When a particle is pulled out of that sea, it becomes real rather than merely virtual—but one can never know in advance just which of the various real states that it *could* become it actually *will* become.) And, to compound this sea of mystery, the various elements of the real states of particles cannot all be measured at the same time: when we measure one of their states (for example, position or energy), another becomes blurred (such as velocity or time of observation).

Perhaps the most remarkable feature of particles is their mutual entanglement. Particles turn out to be highly sociable: once two particles assume a precisely identical state, they remain linked no matter how far they may be from each other. This strange space- and time-transcending connection became apparent when a thought experiment proposed by Einstein with colleagues Boris Podolski and Nathan Rosen (the so-called EPR experiment) was tested by physical instrumentation. The experiment was first performed by French physicist Alain Aspect in the 1980s and has since been replicated in laboratories all over the world. It is important enough to merit deeper acquaintance.

Einstein proposed the experiment in the expectation that it would overcome the limitation on measuring the various states of a particle simultaneously. The idea is to take two particles in a so-called singlet state, where their spins cancel out each other to yield a total spin of zero. Then the particles are allowed to separate and travel a finite distance apart. If the spin states of both particles were then measured, we would know both of the spin states at the same time.

When this experiment is carried out, a strange thing takes place: no matter how far the twin particles are separated, when the spin of one of them is measured, the measurement on the other shows that its spin is precisely the opposite of the spin of the first, as quantum theory requires—even though this was not, and could not have been, determined in advance. It is as if, at the very instant of the measurement, the second particle comes to "know" the state of the first. The information that underlies this strange knowledge appears to be conveyed over any finite distance, and to be conveyed nearly instantly. In Aspect's experiments the speed of its transmission was estimated at less than one billionth of a second, about twenty times faster than the velocity of light in empty space. In a subsequent experiment performed by Nicolas Gisin, it proved to be 20,000 times faster than the speed of light.

Widely reported "teleportation experiments" have shown that such "nonlocal connections" exist not only between individual particles, but also between entire atoms. In the spring of 2004 two teams of physicists, one at the National Institute of Standards in Colorado and the other at the University of Innsbruck in Austria, demonstrated that the quantum state of entire atoms can be teleported by transporting the quantum bits ("qubits") that define the atoms. In the Colorado experiment led by M. D. Barrett, the ground state of beryllium ions was successfully teleported, and in the Innsbruck experiment headed by M. Riebe, the ground and metastable states of magnetically trapped calcium ions were teleported.

The physicists achieved teleportation of a remarkably high fidelity—78% by the Colorado team and 75% by the Innsbruck team—using different techniques, but following the same basic protocol. First two charged atoms labeled A and B were "entangled," creating the instant link observed in the EPR experiment. Then a third atom, labeled P, was prepared by encoding in it a specific coherently superposed quantum state. After atoms A and B were separated, atom A was measured together with the prepared atom P. At that point the internal quantum state of B transformed: it adopted the exact state that was encoded in P. The quantum state of P appears to have been "teleported" to B. Since this destroys the previous quantum state of A by recreating it in P, it recalls science fiction's idea of "beaming" an object (in this case a qubit) from one place to another.

The physical world is strange beyond description, but it is not incomprehensible. Its relevant feature is time- and space-transcending entanglement, known as *nonlocality*. Nonlocality is both a microphysical and a cosmological phenomenon; it involves the very smallest as well as the very largest structures of the universe. English physicist Chris Clarke did not hesitate to affirm that the whole universe is an entangled quantum system—it always has been, and always will be, fully coherent. For U.S. quantum theorist Henry Stapp, the finding of nonlocality is the most profound discovery in all of science. Cosmologists Menas Kafatos and Robert Nadeau, who wrote a book entitled *The Nonlocal Universe,* no doubt agree.

THE STILL MORE AMAZING COHERENCE BETWEEN THE HUMAN BODY, THE MIND, AND THE UNIVERSE

As we have seen, leading biologists and biophysicists find that the living organism is extraordinarily coherent, with all its molecules, cells, and organs linked by quasi-instant multidimensional connections. We have also seen that astronomers and cosmologists discover a comparable level of coherence in the universe as a whole. Are these coherent systems, at widely different scales of magnitude, also coherently linked among themselves? There is mounting evidence that they are. Extraordinary forms of coherence are coming to light between the living organism and the rest of the world.

The Coherence between the Human Body and the Universe

The coherence of the human body with the world around us is shown by its unexpected sensitivity to the substances, forces, and fields that impact it. Critical aspects of this sensitivity exceed the scope of classical biology, including the "synthetic theory" based on the work of Charles Darwin.

According to Darwinists, there should be a full and complete separation of the *genome* (the set of molecules in the DNA that carry the information the cells need to build the whole organism) from the *phenome* (the organism built according to this genetic information). Because of the insulation of the genome from the phenome, genetic information should be fully shielded from external events, including the forces and fields that act on the organism in its environment. Evolution should proceed by a selection from among randomly created genetic variations according to the "fit" of the resulting phenome to its particular environment. In the Darwinist view biological evolution is the product of a two-fold chance: the chance variation of the genome, and the chance fit of the mutant phenome to its environment. To cite the metaphor made popular by Richard Dawkins, evolution is the work of a blind watchmaker: it works by trial and error.

But it now appears that the information contained in the genome of a species is not fully insulated from the phenome, and thus from the world in which the phenome finds itself. The genome turns out to be responsive to external influences. It is indeed thanks to the genome's responsiveness that higher species could evolve from simple origins to the complex multicellular forms they have today—because a fully insulated, randomly mutating genome is not likely to have been able to produce viable organisms within the known time-frames. The "search-space" of possible rearrangements in the genome is so enormous that random processes would take incomparably longer to create viable species than the time that was available on Earth for evolution to branch into the myriad forms that exist today.

The probability that random mutations could produce organisms capable of survival is further reduced by the fact that, as biologist Michael Behe pointed out, most organisms are "irreducibly complex" systems. (A system is said to be irreducibly complex if its parts are interrelated in such a way that removing even one part destroys the function of the whole system.) To mutate an irreducibly complex system into another viable system every part has to be kept in a functional relationship with every other part throughout the process of transformation. Missing but a single part at a single step leads to a dead-end. It is difficult to see how this level of precision could be achieved by genetic variations that are piecemeal and random. Living organisms are so finely tuned to their milieu that any mutation of their genome resulting from random alterations will almost certainly reduce rather than enhance the viability of a species. Random mutations would end up by impairing fitness to the point where the species could no longer survive.

However, the biosphere is populated by a vast number of complex species, the result of a long series of successful genetic mutations. This indicates that mutations in the genome are not always piecemeal and random, but are sometimes massive and systemic. If they are to be successful, the mutating elements of the genome must be highly coordinated among themselves, and must likewise be in harmony with the conditions in which the species finds itself. This suggests that the mutating genome is not fully isolated from the phenome and the environment in which the phenome finds itself. But to claim this is heresy for Darwinism, even in its current form known as the "synthetic theory."

There is solid evidence for the correlation of the genome with the phenome and its milieu. It is both circumstantial and experimental. The circumstantial strand of evidence refers to the time-span of biological evolution. The oldest rocks on this planet date from about four billion years before our time, and the earliest and already highly complex forms of life (blue-green algae and bacteria) are more than 3.7 billion years old. The relatively rapid emergence of these forms of life could not have relied on random mutations alone. The assembly of a self-replicating prokaryote (a primitive non-nucleated cell) involves building a double helix of DNA consisting of some 100,000 nucleotides, with each nucleotide containing an exact arrangement of 30 to 50 atoms, together with a bi-layered skin and the proteins that enable the cell to take in food. This construction requires a long series of reactions, finely coordinated with each other. This level of complexity is not likely to have emerged within the relatively brief period of 300 million years purely through a series of random genetic trials and errors.

Massively coordinated mutations are also needed to produce organisms in the higher stages of evolution. A reptile that can fly is not produced solely by the evolution of feathers, for example. Radical changes in musculature and bone structure are also required, along with a faster metabolism to power sustained flight. It is important to note that each innovation by itself is not likely to offer evolutionary advantage; on the contrary, it is likely to make an organism less fit than the standard form from which it departed. And if so, it would soon be eliminated by natural selection. All this means that evolution by chance mutations is not likely to work. In the words of mathematical physicist Fred Hoyle, its probability is about as likely as a hurricane blowing through a scrap-yard assembling a working airplane.

The experimental strand of evidence concerns the response of the genome to the influences acting on the phenome. Molecular biologists find that mechanical force, or any stress or radiation in the environment, can trigger not just a mutation, but even a global hypermutation. The genome proves to be "fluid"; when confronted with a force or stress, it produces a complex and practically instant series of rearrangements.

A further response of the genome comes to light when electromagnetic or radioactive fields irradiate the organism—this has a direct effect on the structure of its genes. Ingested substances can also produce effects handed down to offspring. Experiments in Japan and the U.S. show that rats develop diabetes when a drug administered in the laboratory damages the insulin-producing cells of their pancreas. The diabetic rats produce offspring in which diabetes arises spontaneously. Even more striking are experiments in which particular genes of a strain of bacteria are rendered defective—for example, genes that enable the bacteria to metabolize lactose. When these bacteria are fed a pure milk diet, some among them mutate back precisely those of their genes that enable them to metabolize it again. Given the complexity of the genome even of bacteria, this response is not likely to occur merely by chance.

It appears that genome, organism, and milieu form a coherent, highly integrated system. It is thanks to this integrated system that living organisms are able to produce offspring that are viable under conditions that would be fatal to the parent organism. That the biosphere is not populated solely by bacteria and blue-green algae, the simplest and most stable kinds of organism, is due in the last analysis to the high level of coherence between genes and organisms, and organisms and environments.

The Coherence between the Human Mind and the Universe

The coherence of our body with the world is fully matched by the coherence of our mind with the world. Of course, our mind is continuously informed of the world outside our body through our eyes, ears, and other sensory organs, so on first sight its coherence in regard to the milieu is not surprising. But the coherence of the human mind with the world is not limited to information conveyed by the sensory organs. There is evidence that when the brain is operating in a coherent way, the mind achieves spontaneous coherence with other minds, and even with other elements in its environment.

A series of experiments carried out by Italian physician and brain researcher Nitamo Montecucco provides impressive testimony to the kind and level of coherence that can obtain between the minds of different individuals. The experiments show that as people enter an altered state of consciousness—in deep meditation or prayer—the electrical activity of the left and right frontal hemispheres of their brain becomes synchronized. Still more remarkably, the electroencephalograph (EEG) patterns of the left and right brain hemispheres of an entire group of persons can become synchronized with one another. In repeated tests up to twelve meditators achieved a 50 to 70 percent synchronization of their EEG waves while sitting in deep meditation in complete silence, with closed eyes and no sensory contact with each other.

Dr. Montecucco summed up the results of his research by distinguishing four states of brain coherence. The lowest state is the "state of disgregation" where coherence is absent. This is a state of no-consciousness, typical of severe bodily degeneration, cerebral death, or deep coma. The next state is the "state of fragmentation," hallmarked by some coherence, but of a low level. This is typical of low awareness, depression, disease, and psychic crisis. The third state is the "state of integration" characterized by a significant level of coherence. It is correlated with high awareness, integrated knowing and experiencing, and psychophysical well-being. The fourth state is the "state of unity." Here the EEG patterns produced by the brain exhibit harmonic relations: the frontal hemispheres of the brain are fully synchronized and move in unison. According to Montecucco this state of the brain and of the consciousness that accompanies it is the state in which great evolutionary leaps occur, in the cognitive, social, as well as spiritual domains.

A growing storehouse of evidence indicates that when the brain functions coherently, consciousness is not limited to the signals conveyed by the senses. This is a surprise to modern people who view extra- or non-sensory perception with skepticism, but it is not surprising for other cultures. Traditional tribes knew and actively used some form of extrasensory perception in their daily life. Shamans and medicine men could induce the altered state of consciousness where spontaneous information-transmission becomes possible, and their spiritual powers appear to have been a consequence of this state. Not just shamans and medicine men; entire clans could enter the states of consciousness in which they could remain in touch even when roaming far from their homestead. Anthropologist Robert Lawlor reports that the aborigines of Australia regularly enter the state of consciousness they call "Dreaming" and in that state come in touch with the spiritual heritage of their ancestors as well as with the mind of their fellow tribesmen. They enter the Dream-state not only in sleep, but also in rituals, during initiation, and in other significant moments of life. A. P. Elkin reported that an aborigine tribesman far from his village might announce that his father has died, or his wife has given birth, or that there is some trouble at home. He is so sure of his facts that he is ready to return at once.

In the roundtable on science and spirituality in part three of this book American anthropologist Ewert Cousins recounts that a similar state of consciousness prevails among the Pottawatomie Indians and the Lakota or Sioux Indians of North America—and that it produces similar effects. Wherever they live, tribes-people appear to be in closer contact with each other and the world around them than modern people.

Sometimes, closer contact with nature saves lives. A dramatic example of this occurred during the Asian tsunami catastrophe of December 2004. Sentinelese and other traditional tribes numbering merely in the hundreds live on the remote Andaman Islands in the Indian Ocean; they have been practically isolated from the rest of the world for thirty- to sixty-thousand years. It was expected that the tsunami would have taken a heavy toll among them, reducing some of their populations nearly to extinction. But this turned out not to be the case. While modern tourists on other shores ignored the impending danger until the murderous waves were literally upon them, the Sentinelese tribes' people left for the highlands in time to escape. They could, of course, have merely followed the birds and animals inland, but it is more likely that they themselves had an inkling of the impending danger, having conserved the sensitivity to nature shared by all primitive people and other living species.

Tribal people, anthropologist Marlo Morgan noted, are able to receive input from their environment as if they have a tiny celestial receiver through which they receive remote messages. Modern people seem to have lost ready access to this celestial receiver, but controlled experiments in parapsychology laboratories show that they did not lose the receiver itself. Under the right conditions most people can become aware of the vague but meaningful images, intuitions, and feelings that testify that they are "in touch" with other people—and with animals, and other aspects of nature—even when these people, animals, and things are beyond the reach of their eye, ear, palate, smell, and touch. This sense of connectedness penetrates to people's everyday consciousness. The *In Our Own Words* opinion survey of the U.S. Fund for Global Awakening reported that 85 percent of Americans agree that "underneath it all we're all connected as one."

However, on the rational level most people in modern society consider extrasensory perception as mere delusion. They do not accept anything as real that is not "manifest"—not literally right at hand. Telepathy and other transpersonal phenomena are admitted only under exceptional conditions. "Twin-pain" (the sensing by one of a pair of identical twins of the pain or trauma of the other) is one such exceptional case: it is recognized as real, for the evidence for it is overwhelming. The transmission of some form of information among identical twins beyond the range of their senses has been studied in detail and it is found to occur repeatedly in about 25 percent of identical twins.

Identical twins are only the top of the tree of emotionally or genetically bonded pairs: some form of spontaneous extrasensory information transmission has been observed among all people who share an emotional or physical bond, such as mothers and children, lovers, long-term couples, even close friends. The basic ability to transmit thoughts and images appears to be given to all people.

The telepathic power of ordinary people is not wishful thinking or a misreading of the results. A whole spectrum of experimental protocols has been developed, ranging from the noise-reduction procedure known as the Ganzfeld technique, to the rigorous "distant mental influence on living systems" (DMILS) method. Explanations in terms of hidden sensory cues, machine bias, cheating by subjects, and experimenter incompetence or error have all been considered, but were found unable to account for a number of statistically significant results. It appears that in altered states of consciousness all or most people possess "paranormal" abilities.

Altered states of consciousness can be produced by, among other things, deep meditation, intense prayer, fasting, rhythmic breathing, rhythmic movements (as in dance), and rapid eye-movements. These states are receptive to information beyond the range of the senses, unlike ordinary waking consciousness in which things that do not fit the dominant conception of the world tend to be repressed.

Transpersonal psychologist Stanislav Grof, who has been investigating altered states of consciousness in a wide range of individuals, claims that everything that a human being can experience in the world through his or her ordinary senses he or she can also experience in the extrasensory mode. And this experiencing can take place no matter how far or how near, or how big or how small are the things that are experienced.

Carl Jung, fascinated with this seemingly esoteric aspect of the psyche, attempted an explanation in terms of what he called the *unus mundi,* a higher or deeper reality that would lie behind both *psyche* and *physis* and connect human minds with each other as well as with nature. He compared unconscious processes in individuals with the myths, legends, and folktales of a variety of cultures at various periods of history and found that the individual's recollections and the collective material contain common elements. According to Jung, these elements make up humankind's "collective unconscious."

The collective unconscious has both a historical aspect and an archetypal aspect. The historical aspect consists of experiences accumulated by human beings throughout history: these experiences have entered, and are conserved in, the collective unconscious of humankind. Archetypes are the dynamic principles that organize their manifold elements. They are irrepresentable in themselves, but have effects that make visualizations possible. Archetypal ideas do not merely repose in the historical dimension of the collective unconscious, but can become part of the waking consciousness of individuals.

The collective unconscious, the same as other transpersonal phenomena, is evidence that our mind is not an isolated entity but is constantly in touch with other minds as well as with the world around us. We are never entirely detached from the outside world; never entirely enclosed within our skin. Our mind and our body resonate with our environment, including other people in our environment. Our mind is coherent with the world, and when we do not repress the intuitions that link us with other people and with nature, we can become aware of our oneness with the universe.

Nearly everything in the world is coherent with nearly every other thing. This is a remarkable finding, for it makes the universe into an interconnected, integral whole. Of course, not every single thing we could name is coherent with every other single thing—that is more than the evidence speaks to. Pebbles, rocks, and mountains are not necessarily and intrinsically coherent with each other and with other things around them. But the particles that make up the atoms that make up the pebbles, the rocks, and the mountains are in fact coherent with each other—and so are the trees that grow on the side of the mountains and the organisms including us humans who walk on the pebbles and the mountains. Universal coherence is a fundamental finding, and it gives us a fundamentally new concept of the nature of reality.

CHAPTER TWO

The Rediscovery of the Akashic Field

We have seen that there is good evidence for the amazing coherence of the universe, of our body and our mind, and of the relation of our body-mind to the universe. But what explanation is there for this coherence? If it is not an incomprehensible mystery, the inscrutable will of a transcendent Creator, there must be something in the natural world that creates it—something that connects and correlates minds and bodies in the biosphere, and quanta and galaxies throughout the cosmos.

Although there have always been signs that something connects things in the natural world, it has not been scientifically recognized until recently. The history of science shows that even fundamental aspects of nature are ignored until innovative scientists piece together the evidence for them. The greatest minds in nineteenth-century physics failed to recognize the existence of the electromagnetic field until Michael Faraday discovered the relationship between electricity and magnetism and James Clerk Maxwell predicted the existence of electromagnetic waves. Today we know that the waves that make up the electromagnetic field exist—we live in a world saturated by them: they make it possible for us to listen to radio, watch television, navigate the Internet, send electronic messages, and use cellular phones. But that there could also be something in nature that connects and creates coherence within and among humans and between humans and the cosmos is only now beginning to be recognized. This recognition will spread because, as we have seen, the evidence for connection and coherence speaks loud and clear:

- Quantum physicists know that there are quasi-instant connections among the particles that populate space-time: every particle that has ever occupied the same quantum state as another particle remains subtly but effectively correlated with it.
- Cosmologists find that the universe is "nonlocal" as a whole, manifesting astonishingly fine-tuned correlations among all its basic constants and dimensions.
- Biologists and biophysicists discover similarly puzzling correlations both within the organism, and between the organism and its environment.
- The connections that come to light in the farther reaches of consciousness research are just as remarkable: they testify to spontaneous interaction between one individual's consciousness and the consciousness of others.

The current findings of connection and coherence ground an important insight. The networks of connection that make for a coherently evolving cosmos, for the entanglement of particles, for the instant connection between organisms and environments, and between the consciousnesses of different and even far removed human beings, suggest that there is more to the universe than matter and energy, space and time. There is also an element that connects and correlates. This element is as much a part of the universe as the electromagnetic, the gravitational, and the nuclear fields. It, too, is a field—a field that is as fundamental as electromagnetism and gravitation, and the fields of the atomic nucleus.

AKASHA AND THE A-FIELD

The idea that there is something in the universe that connects and correlates is not a new discovery; it is essentially a *re*-discovery. As an intuitive insight it is present in all the great cosmologies, most explicitly in Hindu cosmology. There it is known as *Akasha,* the most fundamental of the five elements of the cosmos—the others being *vata* (air), *agni* (fire), *ap* (water), and *prithivi* (earth).

Akasha embraces the properties of all five elements; it is the womb from which everything has emerged and into which everything will ultimately re-descend. It has come into being at the first moments of Creation, appearing in gross form initially as *Sthula,* but soon becoming invisible. Akasha holds the other four elements within itself, but is at the same time outside of them, for it is outside of space and time. It is the birthplace of all things.

In his classic *Raja Yoga,* Swami Vivekananda gave the following account of Akasha:

> It is the omnipresent, all-penetrating existence. Everything that has form, everything that is the result of combination, is evolved out of this Akasha. It is the Akasha that becomes the air, that becomes the liquids, that becomes the solids; it is the Akasha that becomes the sun, the earth, the moon, the stars, the comets; it is the Akasha that becomes the human body, the animal body, the plants, every form that we see, everything that can be sensed, everything that exists. It cannot be perceived; it is so subtle that it is beyond all ordinary perception; it can only be seen when it has become gross, has taken form. At the beginning of creation there is only this Akasha. At the end of the cycle the solid, the liquids, and the gases all melt into the Akasha again, and the next creation similarly proceeds out of this Akasha.

In the traditional conception Akasha is an all-encompassing medium that underlies all things; the medium that *becomes* all things. It is real, but so subtle that it cannot be perceived until it becomes the many things that populate the manifest world. According to the contemporary Indian sage Radja Deekshithar we can experience Akasha: our normal senses do not register it, but we can reach it through spiritual practice. The ancient Rishis reached Akasha through a disciplined, spiritual way of life, and through yoga. That is why they could describe its nature in words, making it an essential element of the philosophy and mythology of India.

In the West it was Aristotle who introduced an Akasha-like fifth element in his philosophy of nature: he called it Aether and said that it fills all of space from the moon upward. In modern times the occult spiritual teacher Madam Blavatsky spoke of Akasha as an emanation of Mulaprakriti, the deeper root of reality. Her student Rudolf Steiner developed this notion as the Akashic Chronicles, a kind of universal "diary" from which humanity can acquire full knowledge of its past and present, as well as future.

A hundred years ago Nikola Tesla—the maverick genius who was the father of modern communication technologies—revived the concept of Akasha within the modern scientific context. He said that there is an "original medium" that fills space, and compared it to Akasha, the light-carrying aether. In his unpublished 1907 paper "Man's greatest achievement" Tesla wrote that this medium, a kind of force field, becomes matter when Prana, cosmic energy, acts on it; when the action ceases, matter vanishes and returns to Akasha. Since this original medium fills all of space, everything that takes place in space can be referred to it. He asserted that this Akashic medium—and *not* the four-dimensional structure put forward at the time by Einstein—is the decisive aspect of space.

However, by the end of the first decade of the twentieth century, physicists adopted Einstein's mathematically elaborated four-dimensional space-time and, with the exception of a few maverick theoreticians, refused to consider any concept of a space-filling ether, medium, or force field. Tesla's insight fell into disrepute, and then into oblivion. But today it is being reconsidered. David Bohm, Harold Puthoff, and a small but growing group of scientists are rediscovering the role of a field that would connect and create coherence—in the cosmos, in the living world, and even in the sphere of mind and consciousness.

This field is not outside of nature: it is the heart of nature. It is the originating ground of all things in the universe, and also their ultimate destination. It is the fundamental medium of the cosmos. It underlies all the particles, and all the forces and fields that govern particles and the systems built of particles, throughout space and time. Science has a name for it: it is the *quantum vacuum* (also known as the "unified vacuum" or "nuether").

THE VACUUM THAT CONNECTS*

The idea of a deeper layer of subtle energy that subtends the physical universe dawned gradually in the course of the twentieth century. At the beginning of the century cosmic space was believed to be filled with an invisible energy field: the luminiferous ether. The ether was said to produce friction when bodies move through it, slowing their motion. But such friction failed to materialize. It was not detected in the beginning of the twentieth century in the famous Michelson-Morley experiments; as a result, the ether was soon removed from the physicists' world picture. The absolute vacuum—space that is truly empty when not occupied by matter—took its place.

The vacuum, however, turned out to be far from empty space. In the "grand unified theories" (GUTs) developed in the second half of the twentieth century the concept of the vacuum transformed from empty space into the cosmic medium known as the zero-point field, or ZPF. (The name of the zero-point field derives from the fact that in this field energies prove to be present even when all classical forms of energy vanish: at the absolute zero of temperature.) In subsequent "super-GUTs" the roots of all of nature's fields and forces are ascribed to the complex and as yet mysterious energy sea known as the "unified vacuum," "quantum vacuum," or "nuether." This energy sea is not merely in outer space; it is everywhere: surrounding and indeed embedding atoms and organisms, as well as stars and galaxies.

The vacuum or nuether is both the originating ground and the

---

*This section gives the gist of the relevant scientific discoveries. Readers interested in greater detail will find it in the author's previous books: in *Science and the Akashic Field* (Rochester, Vt.: Inner Traditions, 2004) and, in a more technical form, in *The Connectivity Hypothesis* (Albany: State University of New York Press, 2003).

ultimate destination of the particles that furnish the universe. These particles first arose from the vacuum in the Big Bang, and they continue to spring forth in a process known as "pair-creation." When energy beyond a very high threshold is injected into the vacuum (for example, in particle accelerators), a particle and an antiparticle spring forth. If they do not meet and annihilate each other, the positive particle establishes itself and the negative particle remains as a hole in the zero-point field of the vacuum.

Positive particles are the building blocks of all the things that populate the observable universe. These particles do not last indefinitely. Eons later, in the final collapse of black holes, what is left of the atoms formed by the particles dies back into the vacuum, and the particles and atoms that had until then populated the universe become virtual again.

The vacuum is not only the womb and the grave of the particles that are the basic units of the observable universe; it is also a cosmic resonance board that continually interacts with them. In the past decades numerous vacuum-interactions have been discovered, including the Casimir force and the Lamb shift.

The Casimir force appears when some wavelengths of the vacuum's energies are excluded between two closely placed metal plates. This reduces the vacuum's energy density with respect to the vacuum energies on the outer side of the plates and the disequilibrium creates a pressure that pushes the plates inward and together. The Lamb shift, in turn, is the frequency shift in the radiation emitted by photons when electrons around the nucleus of an atom leap from one energy state to another. The shift in radiation is due to the photon exchanging energy with the zero-point field of the vacuum.

Another interaction of the vacuum with the observable universe occurs in connection with movement: it is the inertial force we feel as a sudden backward push when we accelerate and a forward push when we decelerate. According to Harold Puthoff, Bernhard Haisch, and Alfonso Rueda, this is an effect of the interaction of the vacuum with the charged particles that make up material objects, including our body. Charged particles such as electrons and quarks are constantly jiggled by the vacuum's zero-point field. If they are at rest, or traveling at constant speeds, the net effect of the jiggling is zero: there is no force acting on the particles. But if a particle is accelerating within the vacuum, it encounters more photons in front of it than behind it. This gives rise to a net force that pushes against the particle. According to Puthoff, Haisch, and Rueda, this is the inertial force.

The theory also claims that a massive body "warps" the fabric of space-time around it. The consequence of the warping is that a particle in the vicinity of the body encounters more photons one its far side than on its near side. This, in turn, produces a net force that pushes the particle toward the body: the force of gravitation.

Although the Puthoff, Haisch, Rueda theory is not accepted by all physicists, its claim that the interaction of particles with the vacuum is responsible for the force of gravitation found independent confirmation in experiments carried out by Hungarian scientists Dezsõ Sarkadi and László Bodonyi. In testing the force of gravity, they have used a large and heavy physical pendulum of vertical bell shape and stiff frame. In numerous experiments the scientists showed that—while in the case of unequal masses the gravitational force satisfies the usual estimates—when the mass of the objects approaches the same value, gravitational attraction between them diminishes. The rate of reduction in the force of gravitation suggests that in the case of objects of precisely equivalent mass, gravity will no longer appear. If so, gravity is neither an intrinsic property of massive objects nor the result of space-time curvature. Rather, it is generated between unequal masses as they interact with the vacuum. In the case of distinctly unequal masses the vacuum is warped, much as an elastic surface is warped by the introduction of uneven weights. The warped vacuum acts on the objects that produced the deformation and creates attraction among them. When the masses are equal, the vacuum is not deformed and the attractive force does not appear.

The quantum vacuum has yet another fundamental effect: it is responsible for the stability of the atoms that are the basic units of all complex forms of matter in the universe. The electrons orbiting the atomic nucleus constantly radiate energy, and they would move progressively closer to the nucleus except for the fact that they absorb energy from the vacuum. The quantum of energy they absorb offsets the energy lost due to their orbital motion and keeps them at a stable distance from the nucleus.

The stability of the Earth's orbit around the Sun is likewise due to interaction with the vacuum. As our planet pursues its orbital path, it loses momentum. Given a constant loss of momentum, the gravitational field of the Sun would overcome the centrifugal force that pushes the Earth around its orbit: our planet would spiral into the Sun. This does not take place because the Earth is constantly deriving energy from the vacuum. Inertia, mass, gravity, and the stability of the atom as well as of the solar system may all be due to vacuum interaction.

Still more surprisingly, matter may owe its very existence to interaction with the vacuum. Experiments reported in 2005 by the Brookhaven National Laboratory show that the vacuum constitutes an extremely dense "gluon field"—that is, a field of the "gluons" (the particles that bind—or "glue"—quarks). Quarks are the fundamental units of protons and neutrons, hence of everything we can think of as matter. This means that the gluon-field of the vacuum is responsible for the enduring presence of all matter in the universe. It is responsible also for the greater part of the mass associated with matter. According to current theory, quarks become up to sixty times more massive in interaction with the quasi-instantly appearing and disappearing knots in the gluon field known as "instantons." More than 95 percent of visible matter appears to owe its existence to the interaction of quarks with instantons in the vacuum's gluon-field.

Vacuum theories are as yet in ferment, but they already tell us that the vacuum is the most fundamental field of the universe. It "glues" quarks, stabilizes atoms and solar systems, and creates mass as well as the forces of inertia and gravitation. It connects all things with all other things. As Puthoff noted, "[O]n the cosmological scale a grand hand-in-glove equilibrium exists between the ever-agitated motion of matter on the quantum level and the surrounding zero-point energy field. One consequence of this is that we are literally, physically, 'in touch' with the rest of the cosmos as we share with remote parts of the universe fluctuating zero-point fields of even cosmological dimensions."

The quantum vacuum as Akashic Record. Even without benefit of the latest vacuum-theories, the experience of Apollo 14 Commander Edgar Mitchell led him to conclude that "the quantum vacuum is the holographic information field that records the historical experience of matter." This may be true, but just how does the vacuum record the "historical experience of matter"?

The detailed elaboration of the answer to this question will constitute the next milestone in the advance of contemporary physics. But in a qualitative yet basically correct way we can already grasp the nature of this process by looking at the sea. The sea, too, records all things in it. And, although its record is not permanent, it connects all things with all other things.

When a ship travels on the sea, waves spread in its wake. These waves affect the motion of other ships—something that has been dramatically brought home to anyone who has ever ventured to sail a small craft in the vicinity of an ocean liner. Vessels that are fully immersed in the sea affect things not only on the surface, but also above and below. A submarine, for example, creates subsurface waves that propagate in every direction. Other submarines—and all fish, whales, or objects in the sea—are exposed to these waves and are affected by them. A second submarine likewise "makes waves," and this affects the first, as well as all other ships, fish, and objects in that part of the sea.

If many things move simultaneously in or on the sea, the water becomes full of waves that intersect and interfere. When on a calm day we view the sea from a height—a coastal hill or an airplane—we can see the traces of ships that passed there hours before. We can also see that the traces intersect and create complex patterns. The modulation of the sea's surface carries information about the ships that created the waves. This has practical applications: one can deduce the location, the speed, and even the tonnage of the vessels by analyzing the wave-interference patterns created by their wake.

As fresh waves superimpose on those already present, the sea becomes more and more modulated: it carries more and more information. On calm days it remains modulated for hours, and sometimes for days. The wave patterns that persist are the memory of the ships that plied the waters. If the patterns were not cancelled by wind, gravity, and shorelines, this memory would persist indefinitely. Of course, wind, gravity, and shorelines do come into play, and sooner or later the sea's memory dissipates. (This, we should note, does not mean that the memory of the *water* disappears. Water has a remarkable capacity to register and conserve information, as indicated among other things by homeopathic remedies that remain effective even when not a single molecule of the original substance remains in a dilution.)

We now apply this process of wave recording and interconnecting to the quantum vacuum. There are analogies, but also differences. In the vacuum there are no things or forces capable of canceling or attenuating the waves. All the particles that make up matter in the universe, and all the particles that are responsible for force, are themselves waves in the vacuum. The particle-waves do not cancel each other. The vacuum is superfluid; motion through it does not produce friction. In this it is like liquid helium cooled to 4°K (four degrees on the Kelvin scale). At this point, near the absolute zero of temperature, helium becomes a frictionless medium in and through which all things move without resistance. In the absence of contrary forces, they would move forever. This is the condition that reigns in the cosmic vacuum—when movement through it is smooth and not approaching the speed of light, it does not produce resistance. We can now understand why we are not ordinarily aware of the vacuum. It does not oppose or hinder our movement, so we no more know it than fish knows water.

Although the effect is subtle, the vacuum interacts with all things: all things "disturb" it. In the language of physics, charged particles "excite (or disturb) the vacuum's ground state." Russian physicists Anatoly Akimov, G. I. Shipov, V. N. Binghi, and co-workers produced a mathematically elaborated theory of this interaction. In their theory the vacuum is a physical substance that extends throughout the universe. This substance registers and transmits the traces of the charged particles that disturb it, for all particles create secondary waves in it. The vortices of these waves link the particles and the objects made up of particles nearly instantly: their group speed is of the order of 10^9 *c*—one billion times the speed of light!

Wave-linking through the vacuum involves a factor that is more subtle than the known forms of energy; it is best called *information*. Information in the vacuum is not a metaphysical feature: it has a scientifically acceptable explanation. It is standard knowledge that particles that have a quantum property known as "spin" also have a magnetic effect; they possess a specific magnetic momentum. Hungarian theoretician László Gazdag pointed out that this magnetic impulse is registered in the vacuum through particle-triggered secondary vortices. (Whether waves create vortices in water or in the vacuum, they consist of nuclei around which circle other elements—H_20 molecules in the case of water, and virtual bosons in the case of the vacuum.) In this way the vortices created by charged particles in the vacuum carry information, much as magnetic impulses do on a computer disk.

The information carried by a vacuum vortex corresponds to the magnetic momentum of the particle that created it: it is information on the state of that particle. The vortices interact: when two or more torsion waves meet, they form an interference pattern. The resulting pattern integrates the strands of information on the particles that created them. Thus the pattern itself carries information on the entire set of the particles that created it.

Vacuum vortices are both supraluminal—faster than the speed of light—and enduring. The "phantoms" generated in the vacuum by charged particles persist even in the absence of the objects that generated them. This was discovered by Vladimir Poponin and his team at the Institute of Biochemical Physics of the Russian Academy of Sciences. Poponin, who has since repeated the experiments at the Heartmath Institute in the U.S., placed a sample of a DNA molecule into a temperature-controlled chamber and subjected it to a laser beam. He found that the electromagnetic field around the chamber exhibits a specific structure, more or less as expected. But he also found that this structure persists long after the DNA itself has been removed from the laser-irradiated chamber: the DNA's imprint in the field is still present. Poponin and collaborators concluded that a new field structure is triggered in the physical vacuum. The phantom effect is a manifestation, they claim, of this hitherto overlooked vacuum substructure.

The conclusion is evident. Vortices in the vacuum record information on the state of the particles that created them, and their interference pattern records information on the ensemble of the particles whose vortices have interfered. In this way the vacuum records and carries information on atoms, molecules, organelles, cells, even on organisms and populations of organisms. There is no evident limit to the information that vacuum-waves could record and conserve. In the final count they can record information on the state of the whole universe.

The interference patterns of propagating waves are holographic patterns in which the information conserved is in a distributed form: present throughout the patterns. This is similar to the way information is present in a hologram created by interfering light waves. In such a hologram the entire image of the recorded object is enfolded throughout the interference pattern. And the whole image reappears when any part of the holographic film or medium is illuminated.

Interfering vacuum-vortices are nature's holograms. They carry information on all the things that excite the vacuum. Since the vacuum is superfluid, this interaction does not produce friction; hence the effects are extremely subtle: they are not registered by most bodily senses or any but the most highly specialized physical instruments. But supersensitive living organisms do register them, as the coherence phenomena reviewed in the foregoing chapter testify.

The interaction of the vacuum with the particles and systems that emerge and evolve in space and time recalls the ancient notion of the Akashic Record, revealing that the Hindu seers were on the right tack. There is a deeper reality in the cosmos, a field that conserves and conveys information and thus connects and correlates. In recognition of this timeless insight, this writer named the re-discovered vacuum-based holofield, "Akashic Field." The Akashic or A-field is a universal field, joining science's four universal fields: the G-field (the gravitational field), the EM-field (the electromagnetic field), and the strong and the weak short-range fields that create attraction and repulsion among the particles of the atomic nucleus.

SCIENCE'S NEW VIEW OF THE WORLD

At the frontiers of the empirical sciences a new view of the world is emerging. It is a view that reenchants the cosmos, for it tells us that the cosmos is an organic, highly coherent, strongly interconnected integral system.

The crucial feature of the emerging view is space- and time-transcending correlation. Space and time do not *separate* things. They *connect* things, for information is conserved and conveyed in nature at all scales of magnitude and in all domains.

The current rediscovery of the Akashic Field as a cosmic holofield reinforces qualitative human experience with quantitative data generated by science's experimental method. The combination of unique personal insight and interpersonally observable and repeatable experience gives us the best assurance we can have that we are on the right tack: an A-field connects organisms and minds in the biosphere, and particles, stars, and galaxies throughout the universe. This transforms a machine-like universe that is blindly groping its way from one phase of its evolution to the next into a whole-system universe that builds on the information it has itself generated.

The world is more like a living organism than a machine. It evolves from the present toward the future on the basis of its evolution from the past to the present. Its logic is the logic of life itself: evolution toward coherence and wholeness, through interconnection and interaction.

Part Two

THE REENCHANTED COSMOS

CHAPTER THREE

The Big Questions

Science's reenchantment of the cosmos gives us new answers to some ancient questions. Although physics, biology, and psychology have currently unseated traditional wisdom about the nature of the world, new knowledge is emerging at the frontiers of these sciences. In the integral world of the newly reenchanted cosmos, even big questions have meaningful answers.

It is time to pose again some of the questions that have baffled the best minds in previous epochs. *What do we know about the origins and the ultimate destination of the universe? And about the origins and the future of life and mind on Earth and in the universe?*

THE BIRTH AND DEATH OF OUR UNIVERSE

People have never ceased to wonder about the origins of the universe. The answers they came up with, though based on intuition, were surprisingly similar. The ancient myths of the world's origins are remarkably consistent: in both East and West they envisage the birth of the world as a stupendous process of self-creation; a process in the course of which the forms and structures of the universe arose by a stepwise progression from primordial unity to manifest diversity. In the Western world the creation story of the Old Testament replaced the previously dominant creation-myths. In the Middle Ages Christians, Muslims, and Jews came to believe that an all-powerful God created the sky above and the Earth below, and all things in between, just the way we find them.

In the nineteenth century the Judeo-Christian story of creation came into conflict with the Darwinian account of evolution. A vivid contrast arose between the view that everything we behold was created intentionally by a divine power and the concept according to which living species evolve from common origins on their own. The contrast fueled endless debates, surviving to this day in the controversy surrounding the teaching of "creationist" vs. "evolutionist" theories in U.S. public schools.

Since the 1930s, the Judeo-Christian creation story has had to contend not only with the Darwinian theory of biological evolution, but also with physical cosmology. If the universe was born in a Big Bang 13.7 billion years ago, the way things are today are not necessarily the way they were created then. How did it come into being 13.7 billion years ago? And was there a world before it came into being?

The best that the Big Bang (BB) theory can say about how the universe came to be is that a random instability took place in a fluctuating cosmic vacuum, the pre-space of the universe. It cannot say either why this instability occurred or why it occurred when it occurred. And it also cannot say why the universe came to be the way it came to be: why it has the remarkable properties it now exhibits. These properties are remarkable indeed, far more remarkable than BB theory can explain. We have already noted the astonishing propensity of the universe to bring forth complex systems such as those associated with life and mind. Let us take a deeper look at the pertinent findings.

The fine-tuning of the basic parameters of the universe involves upward of thirty factors and considerable accuracy. For instance: if the expansion rate of the early universe had been one-billionth less than it was, the universe would have re-collapsed almost immediately; and if it had been one-billionth more, it would have flown apart so fast that it could produce only dilute, cold gases. A similarly small difference in the strength of the electromagnetic field relative to the gravitational field would have prevented the existence of hot and stable stars like the Sun, and thus the evolution of life on planets associated with these stars. If the difference between the mass of the neutron and the proton were not precisely twice the mass of the electron, no substantial chemical reactions could take place, and if the electric charge of electrons and protons did not balance precisely, all configurations of matter would be unstable and the universe would consist of nothing more than radiation and a relatively uniform mixture of gases.

There are further features of this universe that speak to something more than mere serendipity in its evolution. Galaxies formed out of the primordial radiation field when the expanding universe's temperature dropped to 3,000 degrees on the Kelvin scale. At that point the existing protons and electrons integrated into atoms of hydrogen, and these atoms condensed under gravitational pull, producing stellar structures and the giant swirls that made for the birth of galaxies. Calculations indicate that a very large number of atoms would have had to come together to start the formation of galaxies, perhaps of the order of 10^{16} suns. It is by no means clear how this enormous quantity of atoms—equivalent to the mass of 100,000 galaxies—would have come together. Certainly, random fluctuations among individual atoms do not furnish a plausible explanation.

It appears that our universe was born in the heat of an initial explosion that was so finely tuned that galaxies and stellar systems, and planets capable of supporting life, could arise in it. Was the fine-tuned primal explosion merely a lucky fluke? Or was it the work of a transcendental Creator? Creationists believe the latter, while cosmologists have mainly opted for the former. Yet the statistical probabilities speak against the serendipity thesis. According to Roger Penrose's calculations, the probability of hitting on a universe such as ours by a random selection from among the possible alternatives is one in $10^{10^{123}}$. This is an inconceivably large number, and it points to an improbability that hardly merits being taken seriously.

If chance was not the basis of the birth of the universe, was the basis design—perhaps purposeful design? This is the position of theology, but science can now envisage other possibilities. In the last few years "cosmic scenarios" have been elaborated in which the universe in which we find ourselves is not the only universe that exists. A number of mathematically elaborated theories of universe-creation have been put forward—by Andrei Linde, Stephen Hawking, Ilya Prigogine, Fred Hoyle, and Steinhard and Turok, among others—and they make eminent sense. They attest the existence of a meta-universe or *Metaverse,* a vaster and possibly infinite cosmos that was not created in the Big Bang, nor will it come to an end when our universe vanishes in the collapse of the last galaxy-size black holes. It existed before our universe came to be, and will exist after our universe will have left the cosmic stage.

Cosmologies of the Metaverse are in a better position than the Big Bang theory (which is limited to *our* universe) to speak of conditions that reigned before, and will reign after, the lifecycle of our universe. The Big Bang may not have been the sole cause and originator of the universe. This stupendous explosion could have taken place in the womb of an already existing universe, and its coherent features could be the result of conditions in that prior universe.

A Metaverse that pulsates and periodically gives birth to local universes can account for the remarkable coherence we now find in the universe—which is henceforth not "the" but "our" universe. The same way as the genetic code of human parents informs the fetus that develops into their offspring, the Metaverse informed the Big Bang, the otherwise inexplicably precise explosion that gave birth to this astonishingly coherent life-bearing universe. It also gave, gives, and will give birth to other universes, producing periodic universe-creating explosions. Baby-universes evolve as gravitation pulls together their particles in galactic and stellar structures. The kind of evolution we observe in our own universe got under way, and will get under way, time after time.

A number of multi-universe cosmologies are being put forward today. In the scenario of multi-universe evolution advanced by Princeton physicist John Wheeler the expansion of the universe comes to an end, and ultimately the universe collapses back on itself. Following this "Big Crunch," it can explode again, giving rise to another universe. In the quantum uncertainties that dominate the supercrunched state, almost infinite possibilities exist for universe creation. This could account for the fine-tuned features of our universe since, given a sufficiently large number of successive universe-creating oscillations, even the improbable fine-tuning of a universe such as ours has a chance of coming about.

It is also possible that many universes come into being at the same time. This, in turn, is the case if the explosion that gave rise to them was "reticular" —made up of a number of individual regions. In Russian-born cosmologist Andrei Linde's inflation theory, the Big Bang had distinct regions, much like a soap bubble in which smaller bubbles are stuck together. As such a bubble is blown up, the smaller bubbles become separated and each forms a distinct bubble of its own. The bubble universes percolate outward and follow their own evolutionary destiny. Each bubble universe hits on its own set of physical constants, and these may be very different from those of other universes. For example, in some universes gravity may be so strong that they recollapse almost instantly; in others, gravity may be so weak that no stars could form. We happen to live in a bubble tuned in such a way that complex systems, including humans, could evolve in it.

In another scenario baby universes are periodically created in bursts similar to that which brought forth our own universe. The QSSC (Quasi-Steady State Cosmology) advanced by Fred Hoyle, George Burbidge, and J. V. Narlikar postulates that such "matter-creating events" are interspersed throughout the universe. Matter-creating events come about in the strong gravitational fields associated with dense aggregates of preexisting matter, such as in the nuclei of galaxies. The most recent burst occurred some fourteen billion years ago, in excellent agreement with the latest calculation of the age of our universe.

New universes could also be created inside the black holes of one and the same universe. The extreme high densities of these space-time regions represent "singularities" where the known laws of physics do not apply. Stephen Hawking and Alan Guth suggested that under these conditions the black hole's region of space-time detaches itself from the rest and expands to create a universe of its own. The black hole of this universe is the "white hole" of another universe: it is the Big Bang that creates that universe.

Last but not least, we should take note of the cosmology advanced by Paul J. Steinhardt of Princeton and Neil Turok of Cambridge. Their scenario is that the universe (effectively the Metaverse) undergoes an endless sequence of cosmic epochs, each of which begins with a "Bang" and ends in a "Crunch." Each universe-epoch starts with a period of gradual expansion, and is followed by accelerated expansion leading to reversal and the beginning of an epoch of contraction. Steinhardt and Turok estimate that at present we are about fourteen billion years into the current cycle of this universe and at the beginning of a trillion-year period of accelerated expansion. Ultimately our universe will achieve the condition of homogeneity, flatness, and energy under which the next cycle—the next universe in the Metaverse—can begin.

The emerging consensus among leading cosmologists is that local universes evolve, and in time reach the stage where they consist mostly of quasars and black holes. Galaxies collapse on themselves as black holes form at their center (a black hole was recently discovered at the center of our own galaxy), and eventually all galaxies "evaporate" in supergalactic black holes. These may be the womb from which new universes arise. Or, further explosions—"star-bursts"—may occur even in the absence of black holes, and these would be the Big Bangs of subsequent universes.

Whichever multi-universe scenario ultimately stands the test of time, it will have to allow that universes are not entirely disconnected from one another. The connection between universes is through cosmic space: space that endures even as universes come and go. Cosmic space is not empty: it is filled with fluctuating energies that conserve and convey light, force, and information. Space is a cosmic vacuum: the quantum vacuum. The charged particles that arise in each universe "excite" the cosmic vacuum and leave their holographic trace in it. The information "read into" this vacuum is in the form of interfering wavefronts—holograms—and it is not evanescent. It accumulates as fresh waves superpose on those already present. In consequence there is an ongoing transfer of information between the successive universes: the Big Bangs of later universes are informed by the traces of their precursors. The cosmic vacuum is nature's collective memory. It builds throughout the lifecycle of a universe, in universe after universe.

Information that reaches a universe from prior universes—"trans-universe information"—makes for progressive evolution across the cycle of universes. The parameters of later universes become tuned to the processes that unfolded in the earlier universes, and hence they neither collapse back on themselves shortly after their birth nor expand so fast that only a dilute gas of particles survives. They evolve more and more efficiently, and so further and further than their predecessors. There is a cyclically recurring but overall progressive evolution from universe to universe.

As the vacuum records the evolution of each universe, it enables the successive universes to build on the experience of their predecessors. This clarifies the mystery of why our universe is so finely tuned to the evolution of life: it was not the first universe to be born in the Metaverse. Prior universes have tuned the Big Bang that gave rise to it to the conditions they achieved in the course of their own evolution. The fine-tuning of our universe is not a mystery, and it is not a matter of serendipity. *It is a matter of inheritance.*

The Metaverse is not separate and distinct from the universes that arise, die back, and evolve again in it: it is the cyclic universe-creating pulsation itself. When this pulsation itself evolves, it is the Metaverse that evolves. It evolves toward universes that are more and more informed by the traces of prior universes—and thus more and more tuned to the evolution of life and complexity.

Early universes must have been purely *physical* universes. In time, the cycle of universes brought forth physical-*biological* universes, and in further time it will produce physical-biological-*psychological* universes. These mature universes will have super-evolved organisms with super-evolved minds. Consciousness is likely to become increasingly dominant in the cosmos.

Is reaching a physical-biological-psychological universe the deeper meaning and ultimate purpose of the evolution of the Metaverse? Mystical and religious insight supports this thesis. What can science tell us? The ultimate purpose that may underlie the evolution of the cosmos is beyond science's reach. Yet the ensemble of facts before us suggests that the idea that the Metaverse evolves toward life and consciousness, speculative as it is, makes excellent sense.

LIFE ON EARTH AND ITS FUTURE IN THE COSMOS

We come back now from the biggest of the big questions to one that is still big, but somewhat more modest: the question about the origin—and subsequently also the future—of life in the cosmos.

Life on Earth arose nearly four billion years ago, and since then it has been evolving inexorably, if discontinuously, building more and more complex organisms, integrated in more and more complex ecologies. If the laws of nature hold true throughout the universe, wherever conditions suitable for the evolution of life are present, physical, physical-chemical, and ultimately biological and ecological self-organization is bound to get under way. Life is not likely to be limited to our planet.

We now know that conditions suitable for the evolution of life are present in a large number of places. Astronomical spectral analysis reveals a remarkable uniformity in the composition of matter in stars and hence in the planets associated with the stars. The most abundant elements are hydrogen, helium, oxygen, nitrogen, and carbon. Of these, hydrogen, oxygen, nitrogen, and carbon are fundamental constituents of life. Where these occur in the right distribution and energy is available to start chains of reaction, complex compounds result. The active stars with which many planets are associated furnish such energy. It is in the form of ultraviolet light, together with electric discharges, ionizing radiation, and heat.

Nearly four billion years ago photochemical reactions took place in the upper regions of the young Earth's atmosphere, and the reaction products were transferred by convection to the surface of the planet. Electric discharges close to the surface deposited the products in the primeval oceans, where volcanic hot springs supplied further energy. The combination of energy from the Sun with energy stored below the surface catalyzed a series of reactions of which the end products were organic compounds.

There are more than 10^{20} stars in our universe which all generate energy during their active phase. Many, and perhaps most, of them have planets circling them, and on planets with the right chemical and thermal conditions, the kind of self-organization that in time could lead to life is likely to occur.

Just how many potentially life-bearing planets there are is not known with certainty; the estimates vary, but are periodically revised toward higher numbers. Taking a conservative tack, Harvard astronomer Harlow Shapley had originally assumed that only one star in a thousand has planets and that only one of a thousand of these stars has a planet at the right distance from it. He further supposed that only one out of a thousand planets at the right distance is large enough to hold an atmosphere, and that only one in a thousand planets at the right distance and of the right size has the right chemical composition to support life. Even then, Shapley found, there should be at least 100 million planets capable of supporting life in the cosmos.

The astronomer Su-Shu Huang reached an even more optimistic estimate. He considered the time scales of stellar and biological evolution, the habitable zones of planets, and related dynamic factors, and came to the conclusion that no less than 5 percent of all solar systems in the universe should be able to support life. This means not 100 million, but 100 *billion* life-bearing planets. Harrison Brown came up with a bigger number still. He investigated the possibility that many planetlike objects that are not visible exist in the neighborhood of visible stars—perhaps as many as sixty such objects more massive than Mars. In that case, almost every visible star possesses a partially or wholly invisible planetary system. Brown estimated that there are at least 100 billion planetary systems in our own galaxy alone—and there are 100 billion galaxies in this universe!

This highly optimistic estimate has been underscored by a finding of the Hubble Space Telescope in December of 2003. The Space Telescope succeeded in measuring a highly controversial object in an ancient part of our galaxy. It was not known whether this object is a planet or a brown dwarf, but it turned out to be a planet, having two and a half times the mass of Jupiter. Its age is thirteen billion years, which means that it must have formed when the universe was very young, having been in existence barely one billion years. If planets could have formed so early in the life of this universe, many more could have formed in its 13.7 billion year existence than astronomers had previously estimated.

Recent observations indicate that planets keep forming with remarkable speed and abundance to this day. In May of 2004, the Spitzer Space Telescope was trained at a "star nursery" region of the universe known as RCW 49, and in one image uncovered three hundred newborn stars, some not more than one million years old. A closer look at two of the stars showed that they have faint planet-forming disks of dust and gas around them. The astronomers estimated that all 300 might harbor such disks. This is a surprising discovery. If planets form around so many stars, and if they form so soon, they must be highly abundant in the galaxy.

Even if telescopes only allow us to observe a few of the many planets that are likely to exist in the galaxy, the current observations are encouraging. In the last fifteen years twelve hundred Sun-like stars in our vicinity have been scrutinized with ground-based telescopes, and this search has come up with over one hundred extrasolar planets. A particularly promising find was announced in June 2002: the planetary system known as 55 Cancri. It is within hailing distance—forty-one light-years—from us. It appears to have a planet that resembles Jupiter in mass and in regard to orbit. Calculations indicate that 55 Cancri could also have rocky planets much like Mars, Venus, and Earth. Many if not most of such planets could harbor life at some point in their evolution.

Yet planets capable of harboring the more advanced forms of life could be relatively rare. Most of the solar systems in our neighborhood have alien planets in widely eccentric orbits, moving either too far from their host sun to sustain life or moving too close to it. According to Peter Ward and Donald Brownlee, on most planets radiation and heat levels may be so high that the only forms of life that are likely to exist are a variety of bacteria deep in the soil.

Even if they are relatively rare, given the astronomical number of solar systems in this and other galaxies, the number of extrasolar planets capable of supporting advanced forms of life could still be very large. And even if evolution is relatively slow—it could take not millions but a few billion years—complex organisms could still exist on many planets. Some of them could be more evolved than life in our own biosphere, since they could have originated in a more distant past. Astronomers have determined that stars in our region of the galaxy are on the average one billion years older than the Sun. If so, evolution on some of their planets could have had a one-billion-year head start compared with evolution on Earth!

Seeing the cosmos as interconnected and nonlocal requires yet another revision of the estimates of the number of biospheres that could harbor complex forms of life. If the A-field of the vacuum records all things that happen in the universe, it also records and conveys the traces of life. This would make a crucial difference to biological evolution: it would speed up the formation of the molecular arrangements required for self-maintaining and self-replicating organisms. It could spell the difference between evolution reaching merely to the level of bacteria, or attaining the higher stage of multicellular organisms.

Information makes a crucial difference in all processes of progressive development. This is strikingly illustrated in an example given by Fred Hoyle. Take the familiar Rubik's cube, said Hoyle (this is a cube of which the six faces are subdivided into three rows of three color-coded sections each, and each segment can be ordered relative to the others by twisting the cube), and ask a blind person to order the scrambled segments so they all line up. A blind man is handicapped by not knowing whether any twist he gives the cube brings him closer to or farther from his goal. He works by random trial and error, and if he is to be sure that he succeeds, he must try all the possible combinations. These are of the order of 5×10^{18}. If he takes one second to make one move, he will need 5×10^{18} seconds to run through all of them. This, however, is equivalent to 126 billion years—nearly ten times the age of this universe! But if the blind man receives information about the correctness of each move, he can unscramble the cube on the average in 120 moves. That means he will take two minutes, and not as much as 126 billion years to achieve his goal.

Thanks to vacuum-based information, the evolution of advanced forms of life is far more probable than it was hitherto estimated. The availability of interplanetary information means that evolution never starts entirely from scratch, and is never at the mercy of the happy fluke that random mutations come up with organisms that prove viable in their milieu.

Although we do not know of evolved biospheres other than ours, we do know that in our own biosphere life took off surprisingly fast—as already remarked, in a mere 300 million years. This suggests that the early evolution of life on Earth did not rely on chance mutations. Nor did it require the physical importation of the spores of proto-organisms from elsewhere in the solar system, as the "biological seeding" theories of the origins of life suggest. Instead, the chemical soup out of which the first organisms arose was informed with the traces of extraterrestrial life. Life on Earth was not biologically, but *informationally* seeded.

What is the outlook for the long-term future of life on Earth? On this planet life depends on the availability of solar radiation within the critical limits of intensity. These limits will be ultimately breached. In time, the Sun will expand to the red giant stage and engulf all its planets, ours included. Life elsewhere in the universe could continue, but not indefinitely: life in our universe is limited in time. The time-horizons are mind-boggling, but they exist; the physical conditions under which carbon-based life—the only kind we know of—can evolve have temporal bounds.

Whether planets with the precisely suitable thermal and chemical conditions are abundant or rare, such conditions do not last forever. The reason is that sooner or later the active phase of the star whose radiation drives the processes of life comes to an end. Ultimately active stars exhaust their nuclear fuel, and then they either shrink to the white dwarf stage or fly apart in a supernova explosion. The population of active stars is not infinitely replenished. Even if new stars keep forming from interstellar dust, a time must come when no further stars are born.

About 10^{12} (one trillion) years from now, all the stars that remain in this universe will have converted their hydrogen into helium—the main fuel of the supercompacted but still luminous white dwarf state—and they will have also exhausted their supply of helium. In visual observation such stars take on a reddish tint, and then fade from sight altogether. As the energy radiated by active stars is gradually lost, the stars move closer together. The chance of collision among them increases, and the collisions that occur precipitate some stars toward the center of their galaxies and expel others into extragalactic space. As a result, the galaxies diminish in size. Galactic clusters also shrink, and in time both galaxies and galactic clusters implode into black holes. At the time horizon of 10^{34} years, all matter in this universe will be reduced to radiation, positronium (pairs of positrons and electrons), and compacted nuclei in black holes.

Black holes themselves decay and disappear in a process Stephen Hawking calls evaporation. A black hole resulting from the collapse of a galaxy evaporates in 10^{99} years, while a giant black hole containing the mass of a galactic supercluster vanishes in 10^{117} years. (If protons do not decay, this span of time expands to 10^{122} years.) Beyond this humanly inconceivable time horizon, the cosmos contains matter particles only in the form of positronium, neutrinos, and gamma-ray photons.

Although different theories make different postulations about the current state of our universe—as expanding (open), expanding and then contracting (closed), or balanced in a steady state—each implying a different end, in all of them the complex structures required for the known forms of life vanish before matter itself supercrunches, or evaporates.

In the late phases of a *closed universe*—one that ultimately collapses back on itself—the universe's background radiation increases gradually but inexorably, subjecting living organisms to mounting temperatures. The wavelength of radiation contracts from the microwave region into the region of radio waves, and then into the infrared spectrum. When it reaches the visible spectrum, space is lit with an intense light. At that time all life-bearing planets are vaporized, along with every other object in their vicinity.

In an *open universe* that expands indefinitely, life dies out because of cold rather than heat. As galaxies continue to move outward, many active stars complete their natural life cycle before gravitational forces bunch them close enough to create a serious risk of collision. But this does not improve the prospects of life. Sooner or later all the active stars of the universe exhaust their nuclear fuel and then their energy output diminishes. The dying stars either expand to the red giant stage, swallowing up their inner planets, or settle into lower luminosity levels on the way to becoming white dwarfs or neutron stars. At these low energy levels they are too cold to sustain whatever organic life may have evolved on their planets.

A similar scenario holds in a *steady-state universe.* As active stars approach the end of their life cycle, their energy output falls below the threshold where life can be supported. Ultimately a lukewarm, evenly distributed radiation fills space, in a universe where the remnants of matter are random occurrences. This universe is incapable of maintaining the flame of a candle, not to mention the complex irreversible reactions that are the basis of life.

Whether our universe expands and then contracts, expands infinitely, or reaches a steady state, the later stages of its evolution wipe out the known forms of life.

This is a dismal picture, but it is not the whole picture: it doesn't take the Metaverse into account. Life in the cosmos need not end when local universes devolve and disappear. In each universe the forms of life that had evolved leave their traces in the vacuum, and these traces inform the evolution of life in the later universes. Consequently in each successive universe life evolves more and more efficiently. This means that even if it is periodically wiped out, life in the Metaverse evolves further and further.

The evolution of life in the Metaverse is a cyclical process with a learning curve. Each universe starts without life, evolves life when some planets become capable of supporting it, and wipes it out when planetary conditions pass beyond the life-supporting stage. But the quantum vacuum is shared by all universes, and it records and conserves the wave-form traces of all the life that has evolved. As the cosmic vacuum becomes more and more modulated with the traces of life, it becomes more and more favorable to life. It speeds up evolutionary processes wherever they get under way.

Progressive evolution in the Metaverse offers a positive prospect for the future of life in the cosmos. Although life appears and disappears in a cyclical progression, it continues in one universe after another. And it evolves further and further, in universe after universe.

CONSCIOUSNESS AND THE BRAIN

We now lower our sights by another degree and focus on a question that is deep even if not as big as the ones we have just reviewed. This question is of the most immediate relevance to our life. It concerns the most intimately known and at the same time the most mysterious aspect of our existence: our *consciousness*.

Our consciousness accompanies us throughout life, yet we do not know what it is. Is it something produced by the brain? Or is it something different, merely associated with the brain? Common sense opts for the former: on first sight it appears evident that there is consciousness as long as there is brain function. Consciousness, it seems, must be a particularly fancy brain-function.

The thesis that consciousness is a function generated in and by the brain runs into a number of problems, however. First, clinical evidence does not prove that the brain actually *produces* consciousness; it only shows that—with some notable exceptions—it is *associated* with it. Functional MRI (magnetic resonance imaging) and other techniques do show that when particular thought processes occur, they are associated with metabolic changes in specific areas of the brain. But they do *not* show how the cells of the brain that produce proteins and electrical signals would also produce sensations, thoughts, emotions, images, and other elements of the conscious mind—how the brain's network of neurons would produce the qualitative sensations that make up our consciousness. Second, in some well-documented cases (among others, regarding patients suffering cardiac arrest in a hospital) people had detailed and subsequently clearly recalled experiences during the time the EEG wave-patterns of their brain were entirely flat—that is, while their brain was clinically dead. There are also well-documented cases of out-of-body and near-death experiences in which individuals have conscious experiences divorced from their physical brain.

The thesis that the brain actually *produces* consciousness poses a problem that exercises the mind of philosophers. David Chalmers, for example, asks how "something as immaterial as consciousness" can arise from "something as unconscious as matter." How immaterial consciousness could arise out of unconscious matter cannot be answered by empirical brain research, for neurophysiologists and neurosurgeons deal only with gray matter, and matter is not conscious. This, according to Chalmers, is the "hard problem." It is a perplexing problem for, as philosopher

Jerry Fodor pointed out, "nobody has the slightest idea how anything material could be conscious. Nobody even knows what it would be like to have the slightest idea about how anything could be conscious."

Physicist-philosopher Peter Russell responded that Chalmers's problem is not just hard but impossible—at least within the context of the standard world concept of the sciences. Fortunately it does not need to be solved, because this world concept is not the last word. We do not need to explain how unconscious matter generates immaterial consciousness, because consciousness may not be entirely immaterial, and matter may not be entirely unconscious.

We have already seen that matter as such is not a distinct reality in the universe. In his contribution to the roundtable that makes up part three of this book, Russell argues that it is a mistake to think that matter is real. In the final count what appears as matter is a semi-stable bundling of energies within particles, bundled in atoms, further bundled in molecules and crystals, and possibly also in cells and organisms. The cells of the brain—the so-called neurons—are themselves bundlings of molecules, atoms, and particles that are bundles of energy, and the energy configured into particles stems from complex wave-fields extending throughout the cosmos.

There is no way we could dogmatically insist that these energies, and the quanta to which they give rise, lack the qualities we associate with consciousness. Eminent physicists such as Freeman Dyson and philosophers of the stature of Alfred North Whitehead have maintained that even elementary particles are endowed with some form and level of consciousness. Dyson said, "Matter in quantum mechanics is not an inert substance but an active agent, constantly making choices between alternative possibilities. . . . It appears that mind, as manifested by the capacity to make choices, is to some extent inherent in every electron."

If mind is indeed inherent in every electron, the "hard problem" is not a problem. Conscious matter at a lower level of organization (the neurons in the brain) generates conscious matter at a higher level of organization (the brain as a whole). This is because the relatively simple consciousness that is inherent in every neuron in the brain is integrated within the complex structures of the brain and becomes the articulated consciousness we know in our experience.

CONSCIOUSNESS BEYOND THE BRAIN

Consciousness is not the same as the brain, it is merely associated with it. But is consciousness limited to the physical brain with which it is associated, or does it extend beyond it as well?

In the classical view human consciousness is confined to the brain and nervous system. It is the sum total of the sensations, perceptions, volitions, emotions, and intuitions that we experience throughout our life. Information about the outside world enters into this vast and complex stream only through our senses. Neural impulses are conducted by the nervous system from the sensory organs to the brain, where they are decoded and analyzed, and then an item of perception appears in consciousness. Conscious experience is tantamount to looking at the world through five slits in the tower: the eye, the ear, the palate, the nose, and the surface of the skin. Yet this may not be the last word. Could it be possible to attain a broader view—to open the roof to the sky?

We have already noted the remarkable coherence of the human mind with the rest of the world. When we do not repress the information that reaches our consciousness, we become aware of subtle intuitions, images, and sensations that tell us that our brain and consciousness receive spontaneous impressions from our physical and biological, and even from our social and cultural, milieu. In near-death experiences (NDEs), out-of-body experiences (OBEs), past-life experiences, and various mystical and religious experiences, we seem able to perceive things that were not conveyed by our eye, ear, and other bodily senses. In NDEs the brain can be clinically dead, with the EEG "flat," and yet one can have clear and vivid experiences, capable of being recalled in detail. In OBEs one can "see" things from a point in space that is removed from the brain and body, while in mystical and religious transport one has the sense of entering into union with something or someone larger than oneself. Although in some of these experiences the consciousness of individuals is detached from their physical brain, the experiences themselves are vivid and realistic. Those who undergo them seldom doubt that they are real.

Experiences that testify to consciousness extending beyond the physical brain are reported by psychiatrists and psychotherapists as well. They place their patients in an altered state of consciousness and record the impressions that arise in their mind. Transpersonal psychiatrist Stanislav Grof found that in the "experience of dual unity" a person in an altered state of consciousness can experience a loosening and melting of the boundaries of the body ego and a sense of merging with another person in a state of unity and oneness. In the experience of "identification with other persons," merging can be total and complex, involving body image, physical sensations, emotional reactions and attitudes, thought processes, memories, facial expressions, typical gestures and mannerisms, postures, movement, and even the inflection of the voice. In "group identification and group consciousness" one can have a sense of becoming an entire group of people with shared racial, cultural, national, ideological, political, or professional characteristics. It appears that in deeply altered states it is possible to expand one's consciousness to the point where it encompasses the totality of life on the planet. It can extend even to inorganic nature.

Analogous experiences are reported by astronauts. Apollo 14 captain Edgar Mitchell said that in higher states of awareness every cell of the body coherently resonates with "the holographically embedded information in the quantum zero-point energy field." This resonance is logical and understandable. It is conveyed by the vacuum, an Akashic Field that underlies and connects all things, from particles to galaxies. Let us review once again the scientific basis of this remarkable phenomenon.

We have seen that, according to the new physics, the particles and atoms—and the molecules, cells, and organisms—that arise and evolve in space and time emerge from the virtual energy sea known as the quantum vacuum. These things not only originate in the vacuum's energy sea; they continually interact with it. They are dynamic entities that read their traces into the A-field, and through that field enter into interaction with each other.

Interaction through the A-field is not infinite in space and time: it is limited to the cosmic regions where the wave-traces of particles and systems of particles have effectively propagated and created holographic interference patterns. This, in turn, depends on the speed with which the waves propagate. We do not know the exact velocity of vacuum wave-propagation, but it is likely to be high. Vacuum-waves are most probably scalar waves, and scalars propagate at velocities proportional to the density of the medium that carries them. Since the vacuum is an extremely dense virtual-energy field, scalar waves would propagate in it far faster than the speed of light.

Propagating vacuum-waves interconnect vast regions of the galaxy. Within these regions interconnections will persist, for vacuum waves are not subject to attenuation.

Enduring vacuum-interconnection applies to the human brain as well. With the possible exception of OBEs and similar nonordinary-state experiences, all perceptions, feelings, and thought processes have cerebral functions associated with them. These functions have wave-form equivalents, since the brain "makes waves": like other things in space and time, it creates information-carrying vortices. The vortices propagate in our region of the vacuum and interfere with the vortices created by the brains of other people. The resulting interference patterns are complex holograms. The hologram produced by the brain of an individual encounters and interferes with the hologram of other individuals. This creates ever more complex holograms, such as those of a tribe, a community, and a culture. The collective holograms interface and integrate in turn with the superhologram of the species. This is the "collective unconscious": humanity's collective memory.

How can this species-wide memory-bank connect the consciousness of one individual with the consciousness of others? This, too, has a logical explanation. On the principle "like meshes with like," in the collective memory bank we normally access our own hologram. This enables us to read out the information we ourselves have created. This is similar to entering on and reading from the Internet. Reading out what we have read into the A-field is like reading out the materials we have placed on our homepage. If we have the code, we can retrieve all the things we have placed there. In regard to the collective human memory bank we all have the read-out code: it is furnished by our brain. The brain is the complex system that reads the contents of our consciousness into the field, and it is this system that can retrieve it. How this occurs is explained by the interactive process known as "phase conjugation."

Phase conjugation means a relationship between two waves that are equal but opposite in phase to each other. Two equal waves traveling in opposite directions create a single standing wave within which they are in resonance: they are "phase conjugate." Applied to the brain, this means that when the brain receives information from its vacuum-hologram, it is in a phase-conjugate relationship with it. It emits a virtual outgoing wave that is conjugate with the incoming wave from the hologram. In this way the brain can read out the information it has created. This information is conserved in the vacuum hologram, so that our read-out embraces the experiences we have had throughout our life.

The two-way phase-conjugate communication between the brain and the vacuum resolves the puzzle of adequate information storage within the brain. This was a major puzzle, for the brain does not have the capacity to store all the perceptions, sensations, volitions, and emotions experienced by an average individual in a lifetime. Computer scientist Simon Berkovitch calculated that to produce and store a lifetime's experiences the brain would have to carry out 10^{24} operations per second. Dutch neurobiologist Herms Romijn has shown that this would be impossible even if all 100 billion neurons in the brain were involved—and they are not: there are only 20 billion neurons in the cerebral cortex, and many of them have no evident cerebral function. This, however, is not a problem, because long-term memory is not stored *within* the brain; it is *extra-somatic.* The storage of a lifetime's experience is not within the cerebral cortex, but in a holographic field that embeds the brain and the body. This field has staggering storage capacity. In the appropriate phase-conjugate states (for example, in meditation, prayer, contemplation, exaltation, and near the portals of death) it allows all people to recall well nigh everything they have experienced in their life.

This accounts for our own long-term memory, but not for our connection with the consciousness of other people. That phenomenon can also be accounted for, however, by adding a proviso to the read-in/read-out phase-conjugate relationship. We need to allow that this process has sufficient bandwidth to provide access not only to our own hologram, but also to the holograms of others. The "alien" holograms people can occasionally access generally prove to be those of persons with whom they are, or were, physically or emotionally bonded: sisters and brothers, parents and children, spouses and lovers, or simply persons with whom they had evolved ties of friendship and sympathy.

Experience shows that altered states of consciousness are particularly conducive to reading alien holograms—providing access to other people's experiences. This, too, has a logical explanation. It is offered, surprisingly, by fresh light on the brain functions of so-called "savants." Savants—previously often called "idiot savants"—are persons with amazing computational skills or memory, yet subnormal abilities in most other respects; they are often autistic, cut off from normal communication with the outside world. Daniel Tammet, for example, a 26-year old autistic savant, speaks seven languages, can compute cube roots as fast as any calculator, and recalls the constant *pi* to 22,514 decimal places. Such skills occur in about 10 percent of all autistic individuals. Analysis of the pertinent brain functions shows that autistic people preferentially use the right hemisphere of the brain; they experience things in visual terms. The left hemisphere is not dominant and the entire neocortex may be switched off.

The brain, it appears, has two distinct modes of information processing. One is the "classical" mode—slow, linear, and capable of handling only a limited amount of information. It solves problems by using abstract concepts and relies on the connectivity of neural networks in the neocortex. The other is the "quantum processing" mode—extremely rapid, parallel, and capable of handling exponentially more information than the classical mode. This ultrafast mode involves all brain regions, but normally it operates outside conscious awareness. It comes to the fore when classical processing is deactivated.

The brain, as we have already noted, is a macroscopic quantum system. But its nonlocal quantum processing becomes evident only when the left hemisphere does not dominate its information processing, when the neocortex is turned down, or off. This is the case not only in autism, but also in the altered states produced by meditation, prayer, and aesthetic or ecstatic experiences. Because autistic individuals are largely cut off from the world around them, their quantum brain processing turns inward and can produce unusual mental feats. For normal individuals who can communicate with the outside world, quantum brain processing creates contact with the holographic field in which the brain is immersed. Hence, in altered states, the normal brain picks up information from a variety of holograms besides its own.

Transpersonal experiences are not fictional and they are not metaphysical. They are physical, more exactly quantum-physical: readings by the brain of a wide range of holographic information from the A-field.

CONSCIOUSNESS IN THE COSMOS

How widespread is consciousness in the cosmos? In fact, there do not seem to be definite points in the grand architecture of nature where we could say, as of here there is consciousness, and below or above here there are only physical interactions and chemical reactions.

An unbiased view tells us that consciousness is likely to infuse all things in the cosmos, that it is just as fundamental as energy. As energy is bundled into more and more complex configurations, so more and more complex forms of consciousness come to be associated with these configurations. We cannot inspect the consciousness associated with these bundles of energy directly. The evidence for the presence of consciousness in the cosmos is indirect, but it is just as good as evidence for consciousness in our spouse. By direct experience I can know only my own consciousness; everything else is inference.

There are justified as well as unjustified forms of inference. It is justified to hold that only I am conscious and nobody and nothing else is, since in the strict sense I can only experience my own consciousness. In its metaphysical form this is the philosophical doctrine of solipsism (from *solus ipse,* meaning "only the self"), and while it is unarguable, it does not shed light on the prevalence of consciousness in the cosmos. It is also justified to hold that consciousness is associated with my spouse, as well as with my cat and my dog, but then it is not justified to claim that it is not associated with the plants and the insects in my garden, nor indeed with the particles that make up the atoms that make up the molecules that make up my body, and make up all the furnishings of the world. This is neither an outlandish nor a radically new claim: it is contained in the doctrine of *panpsychism.* From Plato and Plotinus to Leibniz and Whitehead, some of the world's greatest philosophers have subscribed to it.

The panpsychist thesis of universal consciousness is not the thesis that consciousness of the kind I myself experience is universal. Rather, it is the thesis that the roots and potentials of the kind of consciousness I experience are inherent in every particle and every atom in the cosmos. It is the thesis that consciousness evolves: it takes on complex forms in complex systems. It is an important thesis, because it questions the primacy of the physical aspect of the cosmos. Ultimately, as we shall discuss in chapter 5, the cosmos must be conceived as neither just physical nor as purely mental. It is best viewed as psychophysical: endowed with equally fundamental mental and physical aspects.

Evidently I—the same as any being in the universe—do not experience the manifestations of the physical and the mental aspects in the same way. Through my brain and nervous system I experience the world around my brain; this is experience of the physical aspect of the body and the universe, that is, "objective" experience. On the other hand when I introspect on my own consciousness, I experience the mental aspect of the universe as it is associated with my brain. I experience myself and all things around me as sensations and perceptions within my stream of consciousness. This is "subjective" experience. I have these distinct objective and subjective experiences, and I cannot reasonably doubt that other beings have them as well. The cat, the tree, even the cells in my body, must experience the world both in terms of "objective" physical-chemical forces and processes, and also as a series of "subjective" sensations, however vague and undifferentiated they may be.

CHAPTER FOUR

The Existential Questions

Can science's newly reenchanted cosmos, with its integral concept of nature, life, and consciousness, shed light on the deep questions of human existence? Can it distinguish between good and evil, and right and wrong? Can it tell us something about the possibility that our consciousness would survive our body—about reincarnation and immortality?

GOOD AND EVIL

The ancient Romans had a saying: *de gustibus non est disputandum*—about matters of taste there is no room for discussion. Taste is subjective: I prefer apples and you prefer oranges; there is no way we could decide who is right. Moral philosophers for the past several hundred years affirmed the same about morals: values, preferences, and beliefs about right and wrong have no objective foundation in the world; they are purely subjective. We cannot apply to observation and reasoning to decide what is good and what is evil and what is right and what is wrong, because from the way something *is,* we cannot derive how it *ought to be.*

However, in science's newly reenchanted cosmos we can discern an objective foundation for morality: we can distinguish between good and evil, and right and wrong. This is because the way things *are* does harbor an indication of how they *ought* to be. Things are not passive and inert, but self- and co-evolving.

In practically all the traditional cultures and religions good and evil have been said to be part of the reality of the world. In the classical doctrines good and evil are "objectified" and often also "personified": they are objective forces embodied in supernatural entities, such as demons, devils, angels, and guardian spirits. But for science this way of conceiving of good and evil does not make sense. The scientific view does not recognize any force or being that could be above and beyond the natural world; all the things we experience either have their roots in our experience of the world or are mere figments of the imagination. This does not mean that the very idea of good and evil has to be purely subjective. There is a sense in which good and evil have their place in the natural universe. They are not supernatural forces and beings—although they may be intuited as such—but events and actions that impact on, and interact with, the evolutionary processes that unfold in nature as well as in the human sphere.

In the scientific context good and evil appear whenever an evolutionary process reaches a level of consciousness where real choice becomes possible. This means reaching the level of consciousness philosophers term "reflexive"—a consciousness that is not just of oneself and of the world, but a consciousness of one's consciousness of oneself and of the world.

Consciousness, we have said, is likely to be universal in the cosmos. But until reflexive consciousness emerges, the presence of consciousness does not warrant qualifying actions as good or evil, for nonreflexive consciousness does not allow autonomous choice. An electron may choose its own quantum state, but it is not likely to choose it in a conscious manner and thus its choice cannot be said to be good or evil: it just is. Even a living organism such as a plant, an insect, a bird, a fish, or an animal does not choose its behavior in a fashion that qualifies as conscious and autonomous. When an impala chooses its strategy of escape from a leopard, and the leopard its strategy of pursuit, they act by combining genetic programming with learning from experience and, as the success of their strategy will disclose, they choose either rightly or wrongly. But they do not choose in a way that is conscious and autonomous, and hence good or evil. That kind of choice is reserved for human beings, and for some other higher primates where choice, informed by a more or less developed reflexive consciousness, becomes relatively autonomous. We humans do not merely act, but act in an autonomous and conscious way; a way that can be considered good or evil in view of its impact on ourselves, and on the world around us.

We can tell if an action merits being considered good or evil in reference to the factor that empowers the evolutionary process—more exactly, in reference to the factor that, if missing, hinders it. This factor is *coherence.* As one particle coheres with another in an atom, one atom with another atom in a molecule, and one molecule, cell, organism, ecology with other molecules, cells, organisms, and ecologies in a biosphere, and, on another level, one solar system, stellar cluster, and galaxy with other solar systems, stellar clusters, and galaxies in the universe, so in a planetary biosphere coherence enables the evolution of higher forms of structure and complexity, accompanied by higher forms of mind and consciousness.

Coherence empowers evolution also in the human world. Coherence in us means health: the optimum functioning of the body. When the body is coherent, its immune system is strong and resistant to disease. When an organ is insufficiently coherent with the rest, it malfunctions: it is diseased. And when a group of cells becomes incoherent it reproduces itself indiscriminately rather than adapting to the organism: it is cancerous.

In turn, coherence *around* us means the integral functioning of the groups and organizations in which we participate: family, community, enterprise, culture, and nation. It also means the integral functioning of nature—a healthy life-supporting environment.

Coherence in us and coherence around us are interrelated and mutually reinforcing. A healthy mind promotes health in the body, and a healthy body promotes mental health. Statistics show that people who are psychologically well adjusted within their family and community are less prone to disease; and those who enjoy good health are less likely to engage in anti-social and anti-ecological behavior.

For us conscious beings capable of real choice, everything we choose to do has a moral dimension. We can lapse into a purposively or inadvertently destructive mode, or we can pull ourselves up by our bootstraps to consciously pursue constructive action. Regardless of our wealth, position, and power, this choice confronts every one of us every day. Everything we do either promotes or counters coherence and thus our and our environment's evolution and development; it is either healthy or unhealthy, and is either constructive or destructive. Moral action promotes, furthers, supports, facilitates, or realizes an aspect or feature of our self- and co-evolution, and immoral action blocks, counters, hinders, and deconstructs an aspect or element of it.

Anti-social and anti-ecological behavior resulting from conscious and voluntary choice deserves to be sanctioned by the laws that regulate behavior in a society, or by the informal but effective moral codes that safeguard the necessary modicum of coherence in every viable community. Significantly and repeatedly incoherence-inducing behavior deserves a rigorous response; it is distinctly anti-social. On the other hand consciously chosen behavior marked by fairness, mutual respect, and solidarity merits being valued and rewarded. Such behavior is cooperative and ecologically responsible; it promotes coherence and hence evolution in the community, in relations with other communities, and in the biosphere.

In an evolving whole-system universe there is a sense in which—for every thing or being—the way it *is* has implications for the way it *ought to be*. The way it ought to be has moral connotations if and only if that thing or being has a real possibility of being the way it ought to be—if it has a choice. As philosophers point out, "ought" implies "can"—it is not reasonable to expect someone to act in a moral way if he or she is not able to do so. In a being who possesses reflective consciousness, this choice means a conscious intention to choose one kind of action in preference to others. For a human being conscious intention decides whether the action is moral and can be rightfully considered good, or it is immoral and deserves to be viewed as evil.

In a subtly interconnected cosmos, where people can access some aspect of each other's consciousness, the role of conscious intention is especially important. One person is likely to intuit coherence-inducing intention in another as the presence of goodness, and incoherence-oriented intentions as the presence of evil. This confers moral responsibility not only on our actions, but also on our intentions. We can bring about goodness in our environment even by tuning our intentions toward coherence, and evil by intending fragmentation, separation, incompatibility, and chaos.

From the intuition of good and evil to their personification is but a small step. We should not be surprised that the great intuitive systems of knowledge—the myths, metaphysics, and religions of humankind—speak of saints and demons, and angels and devils. This does not mean that good and evil exist as supernatural spirits in the world; they exist only in reference to the impact of consciously intended action. "The good" is intention and action that is constructive in regard to the evolutionary process, and "the evil" is intention and action that is destructive.

Creating or contributing to coherence in and around us is not an abstract ideal: living nature is coherent, and so is the universe. There is nothing in the biosphere and nothing in the world at large that would continuously and purposively hinder coherence—unless there are evil civilizations elsewhere in the galaxy. Although in the time span of the past few centuries humanity has become a significant agent of incoherence, creating unnecessary aggression, violence, and irrational fragmentation and polarization, doing so is not part of human nature. Aside from the aberrant psychology of some individuals, it marks only an epoch, perhaps only a brief epoch, in the history of civilization. With more awareness of its nature and its consequences, our generation and the next could pull out of it.

Our ability to consciously intend action that promotes coherence in and around us endows our life with a moral dimension. We not only behold the world; we participate in it—and we can choose to participate in it well. Choosing to do so makes us healthy and whole, creative elements in an integrally evolving universe.

REINCARNATION

The reincarnation of one person's consciousness in the body of another is a perennial belief; it has been an intrinsic part of myth, metaphysics, and philosophy for thousands of years. It is an essential element in Hinduism, Buddhism, Jainism, Sikhism, Zoroastrianism, Tibetan Vajrayana Buddhism, and Taoism. It is present in the belief systems of African tribes, of native Americans and pre-Colombian cultures, of the Hawaiian kahunas, and of the Gauls and Druids. It was adopted by the Essenes, the Pharisees, the Karaites, and other Jewish tribes and groups; it remains an important element in the Kabbalah. In ancient Greece the Pythagoreans and the Orphics subscribed to it. Plato spoke of "metempsychosis" (the transmigration of the psyche) in many of his famous dialogues—Phaedo, Phaedrus, Republic, and Timaeus—Julius Caesar mentioned it as a doctrine held by the Celts, and Roman historians noted that it was shared by the Germanic people. Reincarnation was part of Christian doctrine as well, until it was removed by the Second Council of Constantinople in 553 C.E.

Belief in reincarnation is backed up by noteworthy personal experience, recurring in all societies and historical epochs. People in all walks of life and levels of education seem occasionally to receive impressions and images about sites, persons, and events they have not and could not have encountered in their present life. For the most part they believe that these are memories stemming from a previous lifetime.

More systematic evidence for reincarnation comes from the records of psychotherapists. Therapists who do regression analysis place their patients in an altered state of consciousness—breathing exercises, rapid eye movements, or simple suggestion are usually sufficient for this—and take them back from their current experiences to the experiences of the past. It turns out that they can move the patients back to early childhood, infancy, and even birth. Experiences that seem to be those of gestation in the womb may surface as well.

Interestingly, and at first unexpectedly, some therapists find that they can take their patients back still further. After an interval of apparent darkness and stillness, experiences of other places and other times surface. The patients not only recount these experiences as though they were incidents in a novel they have read or a film they have seen, but seem actually to re-live them. As Grof's records testify, they *become* the person they experience, even to the inflection of voice, the language (which may be one the patient has never known in his or her present lifetime), and, if the experience is of infancy, the involuntary muscle reflexes that characterize infants.

Ian Stevenson of the University of South Carolina investigated the past-life experiences of children. During more than three decades Stevenson interviewed thousands of children, in both the West and the East. He found that from the age of two or three, when they begin to verbalize their impressions, until the age of five or six, many children report identification with people they have not seen or even heard of in their young lives. Often these reports can be verified as the experience of a person who had lived previously, and whose death matches the impressions reported by the child. The case of Parmod Sharma, one of twenty subjects described in one of Stevenson's books, is particularly impressive and merits extensive citation.

> Parmod Sharma was born on October 11, 1944, in Bisauli, India. His father was Professor Bankeybehary Lal Sharma, a Sanskrit scholar at a nearby college. When Parmod was about two and a half, he began telling his mother not to cook meals for him any more, because he had a wife in Moradabad who could cook. Moradabad was a town about a ninety miles northeast of Bisauli. Between the ages of three and four, he began to speak in detail of his life there. He described several businesses he had owned and operated with other members of his family. He particularly spoke of a shop that manufactured and sold biscuits and soda water, calling it "Mohan Brothers." He insisted that he was one of the Mohan brothers and that he also had a business in Saharanpur, a town about a hundred miles north of Moradabad. . . . He provided many details about his shop, including its size and location in Moradabad, what was sold there, and his activities connected to it, such as his business trips to New Delhi. He even complained to his parents about the less prosperous financial situation of their home compared to what he was used to as a successful merchant. . . . Parmod had a strong distaste for curd, which is quite unusual for an Indian child, and on one occasion even

advised his father against eating it, saying that it was dangerous. Parmod said that in his other life he had become seriously ill after eating too much curd one day. He had an equally strong dislike for being submerged in water, which might relate to his report that he had previously "died in a bathtub." Parmod said that he had been married and had five children—four sons and one daughter. He was anxious to see his family again and frequently begged his parents to take him back to Moradabad to visit them. His family always refused the request, though his mother did get him to begin school by promising to take him to Moradabad when he had learned to read. . . . News of Parmod's statements eventually reached the ears of a family in Moradabad named Mehra, which fit many of the details of his story. The brothers of this family owned several businesses in Moradabad including a biscuit and soda shop named "Mohan Brothers." The shop had been named after the eldest brother, Mohan Mehra, and had originally been called "Mohan and Brothers." This was later shortened to "Mohan Brothers." This shop had been started and managed by Parmanand Mehra until his untimely death on May 9, 1943, eighteen months before Parmod was born.

Parmanand had gorged himself on curd, one of his favorite foods, at a wedding feast, and had subsequently developed a chronic gastrointestinal illness followed later by appendicitis and peritonitis from which he died. Two or three days before his death, he had insisted, against his family's advice, on eating more curd, saying that he might not have another chance to enjoy it. Parmanand had blamed his illness and impending death on overeating curd. As part of his therapy during his appendicitis, Parmanand had tried a series of naturopathic bath treatments. While he had not in fact died in a bathtub, he had been given a bath immediately prior to his death. Parmanand left a widow and five children—four sons and one daughter.

In the summer of 1949, the Mehra family decided to make the trip to Bisauli to meet Parmod, who was a little under five years old at the time. When they arrived, however, Parmod was away and no contact was made. Not long thereafter, Parmod's father took him to Moradabad to explore his son's compelling remembrances first hand. Among those who met Parmod at the railway station was Parmanand's cousin, Sri Karam Chand Mehra, who had been quite close to Parmanand. Parmod threw his arm around him weeping, calling him "older brother" and saying "I am Parmanand." Parmod had not used the name Parmanand before this meeting. It is for Indians common to call a cousin "brother" if the relationship is a close one, as was the case for Parmanand and Karam. . . . Parmod then proceeded to find his way to the "Mohan Brothers" shop on his own, giving instructions to the driver of the carriage, which brought them from the station. Entering the shop, he complained that "his" special seat had been changed. In India it is customary for the owner of a business to have an enclosed seat—a gaddi—located near the front of the store where he can greet customers and direct business. The location of Parmanand's gaddi had in fact been changed some time after his death. Once inside, Parmod asked: "Who is looking after the bakery and soda water factory?" This had been Parmanand's responsibility. The complicated machine, which manufactured the soda water, had been secretly disabled in order to test Parmod. However, when it was shown to him, Parmod knew exactly how it worked. Without any assistance, he located the disconnected hose and gave instructions for its repair.

Later at Parmanand's home, Parmod recognized the room where Parmanand had slept and commented on a room screen that he correctly observed had not been there in Parmanand's day. He also identified a particular cupboard that Parmanand had kept his things in, as well as a special low table, which had also been his. "This is the one I used to use for my meals," he said. When Parmanand's mother entered the room, he immediately recognized her as "Mother" before anyone else present was able to say anything. He also correctly identified Parmanand's wife, acting somewhat embarrassed in front of her. She was, after all, a full grown woman and he was only five, though apparently possessing

> at least some of the feelings of an adult husband. When they were alone, he said to her: "I have come, but you have not fixed bindi," referring to the red dot worn on the forehead by Hindu wives. He also reproached her for wearing a white sari, the appropriate dress for a Hindu widow, instead of the colored sari worn by wives.

Stories such as these are clearly not products of the unfettered imagination. They are veridical in the sense that those who recount them "remember" them. But what these memories signify may not be what most people believe they do. The experiences themselves only indicate that many children, and some adults in altered states of consciousness, can access the consciousness of other people, whether these people are near or far, living or dead. Having these experiences does not warrant that the people who have them are the flesh-and-blood reincarnation of those whose experiences they recall.

In light of the coherence and interconnection we find in the domains of nature as well as in the sphere of consciousness, the more modest yet still astounding explanation of past-life experiences is that they constitute access to the holographic vacuum-traces of another person's consciousness. If we do not distinguish an "alien" hologram from our own, we re-live another person's consciousness *as* our own. The person whose experiences we re-live may live in another place, or could have lived at another time. The holograms that carry their lifetime experiences are distributed in space and do not phase out in time.

Those who undergo past-life experiences cannot, and ordinarily do not, distinguish other people's holograms from their own. For them, long-term personal memory merges imperceptibly into transpersonal memory.

IMMORTALITY

If all that we experience becomes part of the collective memory-store of our species, there is a sense in which the consciousness of each of us survives our brain and body. The question is, how does a person's consciousness survive: as part of a recollection stemming seemingly from one's own past life, or as the integral "spirit" or "soul" of a person who is no longer alive, yet capable of communicating with others who are? The former kind of experience furnishes the phenomena traditionally interpreted as evidence for reincarnation, while the latter suggests something entirely different: it suggests *personal immortality*. This is a distinct phenomenon in which ongoing consciousness appears to someone living as another person also still living, but living elsewhere, perhaps on another plane of reality or in another dimension.

Experiences of this kind have been reported by various sources, traditional as well as modern. NDE researcher Raymond Moody collected a wide variety of "visionary encounters with departed loved ones," and mediums such as James Van Praagh, John Edward, and George Anderson have mediated contact with thousands of deceased by describing the impressions they receive from them.

Encounters with deceased persons who were dear to them often occur in people who are themselves close to death: if they recover consciousness they often speak of meeting past members of their family. Dutch cardiologist Pim van Lommel, who has done extensive research on NDEs of patients in intensive care units, came across a number of encounters of this kind. A particularly interesting encounter was recounted by one of his patients who said that during his cardiac arrest he saw his deceased grandmother. He also saw a man who looked at him lovingly but whom he did not know. "More than ten years later," the patient reported, "at my mother's deathbed, she confessed to me that I had been born out of an extramarital relationship; my father, being a Jewish man, had been deported and killed during the second World War, and my mother showed me his picture. The unknown man that I had seen more than ten years before during my NDE turned out to be my biological father."

Encounters with deceased persons can also be experienced by people in good health. A striking experience was reported by Sabine Wagenseil, a German management consultant who, during Stanislav Grof's holotropic breathing seminars in Switzerland, made the acquaintance of Wolfgang Abt, a Benedictine priest. Father Abt, who also participated in the seminars, died unexpectedly of a heart attack on the thirteenth of January 2001. On the first of February Sabine woke around four AM from a dreamful sleep to hear Father Abt speaking to her. After that he spoke to Sabine regularly, announcing his presence by a touch-like sensation on her shoulders or hands. The sessions took place as agreed between Sabine and the deceased, mostly between six and eight in the morning when Sabine was in a meditative state and there were no noises or other disturbing factors in the house. The series of sessions ended on the second of December of the same year, when Father Abt announced that he would contact her less frequently, since the message he wanted to convey had been completed and he was facing new tasks having to do with peace in the world.

The message reported by Sabine Wagenseil was published in 2002 in the reputed German frontier-science magazine *Grenzgebiete der Wissenschaft*. It is in the form of written transcripts of the voice Sabine heard internally—a voice that she was at first reluctant to accept as real and belonging to Father Abt but that she was forced to accept as such when it became clear that it contained material she could not have come up with herself. The voice said that the dead are not "gone" but continue to be present, albeit in another frequency domain. All is vibration in the world, but living and dead are separated by a difference in frequencies. It also said that the dead can perceive ("feel") each other even though they do not have a physical body, and they can hear and see the living, and are indeed constantly concerned with and about them. They are saddened that the living do not recognize them but consider them gone and inaccessible. But, the voice continued, through meditation, contemplation, and prayer, the living can become more and more "transparent," more "open to God," until the time comes when they can apprehend the dead despite the difference in their frequency.

In addition to spontaneous contact, there appear to be ways to communicate with the recently deceased in a purposefully designed way. One variety of communication is, curiously enough, instrumental: it makes use of radio, television, and now the computer. The idea goes back to the 1920s, when Thomas Edison attempted to construct a machine for this purpose. In an interview published on October 30, 1920, in *Scientific American,* he said,

> If our personality survives, then it is strictly logical or scientific to assume that it retains memory, intellect, and other faculties and knowledge that we acquire on this Earth. Therefore, if personality exists after what we call death, it's reasonable to conclude that those who leave the Earth would like to communicate with those they have left here. I am inclined to believe that our personality hereafter will be able to affect matter. If this reasoning be correct, then, if we can evolve an instrument so delicate as to be affected, or moved, or manipulated by our personality as it survives in the next life, such an instrument, when made available, ought to record something.

A variety of such instruments have been developed since the 1950s, making use of tape recorders, as well as radios and television receivers. First these instruments are allowed to produce the indeterminate chaos that comes through as mere static on the air and "snow" on TV screens. Then, after careful preparation, which involves the fine-tuning of the observer's consciousness, voices and images appear, which can be recorded and reproduced. A group in Europe, the Transcommunication Study Circle of Luxembourg (CETL) developed a way to inscribe such messages on the hard disks of computers as well, and in the 1990s recorded tens of thousands of voices and images. The messages describe in some detail a spirit-world inhabited by the mind or soul of the deceased until they either reincarnate, or move to higher spheres that are no longer within reach of communication. They are varied, but unanimous in agreeing that souls or spirits are not separate from each other: in the last count, "we are all one." They also affirm that there is a band of energy or consciousness that surrounds the planet, corresponding to the ancient intuition of the Akashic Record of every event that has ever happened in the past, and will or could happen in the future.

Another form of after-death communication makes use of an induced altered state of consciousness in the receiver. Allan Botkin, a qualified psychotherapist, head of the Center for Grief and Traumatic Loss in Libertyville, Illinois, and colleagues claim to have successfully induced after-death communications (ADCs) in nearly three thousand patients. They report that ADCs can be induced in about 98 percent of the people who try them. Usually the experience comes about rapidly, almost always in a single session. It is not limited or altered by the grief of the subject or his or her relationship to the deceased. It also does not matter what the experiencer believed prior to undergoing the experience; he could have been deeply religious, agnostic, or even a convinced atheist. Typically, the experience is clear, vivid, and thoroughly convincing. The therapists hear their patients describe communication with the deceased person, hear them insist that their reconnection is real, and watch repeatedly as the patients move almost instantly from an emotional state of grieving to a state of relief and elation.

ADCs can also occur in the absence of a personal relationship with the deceased, for example, in combat veterans who feel grief for an anonymous enemy soldier they have killed. ADCs can also occur without guidance by a psychotherapist. Indeed, as Dr. Botkin reports, leading the experiencing subject actually inhibits the unfolding of the experience—it is sufficient that the therapist induces the mental state necessary for the experience to occur. This state is an altered state of consciousness brought about by means of a series of rapid eye movements. Known as "sensory desensitization and reprocessing," it produces a receptive state in which people are open to bringing to their normal state of consciousness a wide range of the impressions that flow into their mind.

Clearly, ADCs have remarkable therapeutic value. Are they merely grief-induced delusions? Botkin argues that they are not: they do not fit any of the known categories of hallucinations. If so, are they real: do the subjects actually encounter the deceased for whom they are grieving? That would suggest that the deceased still exists in some way, perhaps in another dimension of reality. This would be true immortality: the survival of the person after the physical demise of the body. Botkin maintains that this may be the case. In a letter to the author in February 2005, he said that, although he does not say so in his own book, having read the author's *Science and the Akashic Field,* he now leans toward believing that individual consciousness does survive death. "It seems reasonable to me," he wrote, "that all of the bits of information associated with individual consciousness remain highly 'entangled' or 'correlated' even after physical death, and that this continued entanglement may account for individual consciousness remaining intact and active."

This conclusion is backed by individual experimentation with altered states of consciousness. Chris Bache, a professor of philosophy currently in Ohio, carried out one of the most sustained and courageous explorations of deeply altered states ever attempted. The following is part of a letter he wrote to the author in July 2005, just as the present book was nearing completion:

> In *Science and the Akashic Field* you essentially dissolve the question of extended individual identity by pointing to the A-field's holographic retention of everything thought, felt, and sensed. You reinterpret the data of ADCs and reincarnation by pointing to the A-field's capacity to generate all experiences of apparent postmortem existence and pre-birth memory, thus making it unnecessary to postulate intermediate structures such as the soul or the continuity of individual consciousness in rebirth.
>
> I am receptive to your proposal both from theory and from my own transpersonal experience. As one "drops" into the vacuum layer by layer, structures that had appeared "real" from the perspective of space-time become transparent to the underlying ground. Not only ego but soul too, traditionally conceived, becomes permeable and fleeting, simply another intermediate organization of experience in an open, unbounded system of energy and information. All patterns of life including collective, archetypal, even cosmic structures can become transparent, sunyata, empty of self-existence. And yet there is another piece to the puzzle, I think, that this account leaves out.
>
> From twenty years of exploring nonordinary states of consciousness, I have taken away a small number of fundamental convictions. One of these convictions is what I call the principle of accumulation, that is, the universe's interest in and penchant for accumulating

experience and with experience insight, understanding, and new capacities. As more experience and insight is gathered and integrated, structures emerge to hold, crystallize, give form to, and further empower these capacities. . . .

In my inner work, wherever I touched my life in nonordinary states, it broke open to reveal a tapestry of collective threads. I could not find any part of "my" existence that was not part of the larger tapestry of life. And yet, over the course of years, things happened that seemed to suggest that something was being birthed in these experiences that would endure beyond any frame of reference previously imaginable to me, beyond egoic existence, beyond any space-time structure altogether. I found it necessary to affirm the emergence of a new and higher form of individuality being generated by the universe's relentless accumulation of experience—both accumulation through many cycles of reincarnation, and through the systematic integration into one point of awareness of vast territories of transpersonal experience. Furthermore, what was happening to me seemed to be simply an instance of a larger pattern operating within the universe at large. In the end, it was the Metaverse evolving itself in and through the universe. The Metaverse was birthing itself into new forms, and at critical thresholds these forms endured. They were not dissolved by death. . . .

Bache concluded,

The universe relentlessly generates, collects, and integrates experience. In time, self-emergence yields systems strong enough to begin to form continuities across multiple incarnations. At first fragmented and disjointed, these lives eventually come together to form more integrated wholes. This process of accumulation and integration reaches a bifurcation point where something comes into being that had not previously existed. I experienced this birth as an explosion of brilliant, luminous, extremely dense, exquisitely clear, crystalline diamond light. . . . This birth seemed to be made pos-

> sible by the accumulation of learning generated by a maturation of consciousness across many lifetimes. Not reincarnation of a soul, as if the soul existed prior to this point, but a reincarnation that gives birth to the soul.

We do not know at this time how individuality—the factor traditionally identified as the soul—could be preserved after death. We do not know how the fundamental information-field of the cosmos could conserve the traces of the body and mind of a person so they can be re-experienced as the body and mind of an individual person—experienced not only as images from the lifetime of that person but as an individual who continues to communicate even when he or she is no longer alive. This, however, seems to be the case. Apparently, the holographic traces of a lifetime are not merely conserved in the A-field but, at least for a time, are capable of autonomous development in the absence of the brain and the body that produced them. In the process of enduring interaction of a human consciousness with the A-field, something akin to the traditional notion of an individual soul appears to be generated.

This is as far as we can now go in explaining the phenomenon of survival beyond death. But the book is not closed on this burning existential question, and the promise of further insight is real. Researching the phenomena of a surviving soul or individuality and relating it to information conserved in nature is one of the most promising, and certainly the most challenging, of the many tasks of fundamental research that await physicists, biologists, and students of the human psyche in coming years.

CHAPTER FIVE

The Integral Vision of Reality

After our explorations of the big questions of universe, life, and consciousness, and the existential questions of morality, reincarnation, and immortality, it is time to get down to the most fundamental question of all. *What is the real nature of the world we live in?* If we do not *create* our reality but just *experience* it, there should be "something there" even when we do not experience it. This something is the reality of the world. It is not an entirely independent reality, for the very fact of our experiencing it enters into it. Our experience of the world interacts with the world, but it does not create it. We can disagree with the radical interpretations of quantum physics and agree with the realism of Einstein: the fact that I see a mouse when I look at it does not mean that there is no mouse when I do not look at it.

But just what is the reality that underlies our experience of the world? The reality of a reenchanted cosmos must be a different reality from the reality of a materialistic and mechanistic universe. There, reality is basically matter moving about in space and time. What is reality in avant-garde science's current concept of the world?

The first thing we should note is that our experience of the world does not give us reality in its total and pristine purity. Reality may be like a precious diamond: it may have many facets. In human experience we get only one facet—the humanly experience-able facet. It may be that there is an absolute, transcendent reality that is deeper or higher than this facet. But whether there is such a reality or not, it is beyond the reach of science, for in the final count science must be based on experience. If so, we should rephrase our query to: *What is the scientifically knowable facet of reality?*

Even if science cannot grasp a transcendent ultimate reality, there is more to the humanly accessible facet than what we can see, hear, taste, smell, and touch. Scientists, like everyone else, inspect the world with their eyes and ears but, unlike other people, they also inspect it through instruments that extend the power of their senses. They measure and record what they experience and compute the results—but they do not reduce everything to it, because what scientists experience at any one time is not always and necessarily consistent with what they have experienced at other times. Sometimes, as in reading the results of new experiments, it may even be contradictory. Then scientists are puzzled, but they do not give up. They create theories that account for the surprising features of experience by going beyond the directly experience-able domain, postulating such seemingly esoteric entities as quarks and black holes, and such abstruse forces as probability waves and quantum gravity. The non-experienced and non-experience-able domains of the world are held to be just as real as those that are experienced, for they render the sphere of experience consistent and coherent.

Belief in the consistency and coherence of the world is a fundamental pillar of science. Scientists can accept that nature has indeterminate and even chaotic aspects, but cannot accept that it is entirely incoherent and haphazard.

Combining the immediately sensed domains of reality with the domains one assumes lie beyond is a tried and tested way of proceeding, for the world of direct experience is not a humanly meaningful world. Animals live in such a world, but humans have gone beyond it from the very beginning of history. In times past we populated the sky above and the earth below with gods and devils, angels and demons, and in modern times we are populating it with microparticles and supergalactic clusters, with multi-dimensional space-time and a plethora of fields and forces that cannot be visualized.

Science's "leap beyond" the directly experience-able domain of reality is carefully reasoned and thoroughly tested. Even then, science's concept of reality is not necessarily the final and ultimate truth, for every scientific theory must remain open to correction and even substitution. In fact, scientific theories are constantly tested for coherence and consistency both with experience, and with previous knowledge about experience. For this reason, even if the reality disclosed at the leading edge of contemporary science is not the last word, it is likely to be a good approximation. It merits being known—and debated and pondered, for science's new vision of reality is an integral vision, pregnant with meaning for our life and times.

THE CONCEPT OF PHYSICAL REALITY

We begin with science's concept of physical reality. Like all theories of science, this concept is open to revision. Yet the current concept may not actually be in *need* of revision—at least in regard to what it is *not*. We can affirm with a high degree of confidence that physical reality is not confined to matter in space and time. Ours is not a materialistic universe where matter moves about in neutral space, governed by simple rules of cause and effect as in a machine. Instead, our best insight is that this is an evolving, instantly and enduringly interconnected fundamentally integral universe, embedded in a dynamic and physically real cosmic medium. The fundamental element of this integral universe is "physical space-time," more exactly, the physical vacuum that extends throughout space and time.

The realization that the vacuum that subtends all things in the universe is the fundamental medium of physical reality may be the "bright new idea" that physicists believe is required to get them out of their current impasse—the impasse of trying to evolve so-called "string theory" into the master theory known as the Theory of Everything. The "ToE" is expected to give a single, unified explanation—at best a simple and basic equation, similar in scope to Einstein's $E=mc^2$—of all the laws and constants of physical nature, including the way space and time are generated.

In string theory (which is the theory physicists look to for creating a ToE), vibrating strings replace the concept of particles. Strings vibrate at different frequencies, and each frequency defines a corresponding kind of particle: one "note" on the string makes for an electron, another for a neutron, still others make for bosons and gravitons, the particles that carry the forces of nature.

The impasse physicists face now is a kind of embarrassment of riches: there are far too many solutions to the equations of string theory—perhaps of the order of 10^{500}! Worse than that, each solution describes a different universe: a universe with different laws and properties. In some light could only travel short distances, in others there would be many more quarks than the six our universe seems to have, in yet others space would have nine dimensions, and in still others the whole universe would be reduced to microscopic dimensions. One way out of this quandary is to assume that all possible universes co-exist, although we apprehend only one among them. Another way is to assume that our own universe has all these almost infinite number of faces, even if we experience just one.

But string theory faces a problem that is still more vexing: it requires space and time to contain the strings, but it cannot show how the strings would generate space and time. Space and time must preexist independently, and if string theory does not account for space-time, it is not a true ToE. It turns out that another revolutionary theory can do better in this regard: the theory of "loop quantum gravity." Here space and time are woven from a network of nodes and links that connect all points, even distant ones. These links explain how space-time is generated, and also account for "action-at-a-distance," that is, for the entanglement that underlies nonlocality. Despite the great expectations attached to it, string theory is not the final word: it requires further theories or assumptions if it is to make sense of physical reality.

The new fundamental idea that more and more physicists now believe is needed to achieve a comprehensive understanding of the nature of physical reality is before our eyes, and it is not actually new. *It is to consider space as the fundamental medium of the cosmos.* The great pioneers of modern physics have already entertained this notion. In the nineteenth century William Clifford, the creator of the modern Clifford algebras, affirmed that small portions of space are analogous to little hills on a surface which is, on average, flat; the ordinary laws of geometry do not hold for them. The property of space to be curved or distorted is continually being passed on from one portion of space to another after the manner of a wave. This variation in the curvature of space is what really happens when matter moves. Thus in the physical world nothing else takes place but this wavelike variation.

Half a century later, in a paper entitled "The Concept of Space," Einstein wrote, "We have now come to the conclusion that space is the primary thing and matter only secondary; we may say that space, in revenge for its former inferior position, is now eating up matter." A few years following the publication of Einstein's thought, Schrödinger restated the basic insight. "What we observe as material bodies and forces," he noted, "are nothing but shapes and variations in the structure of space."

Can the structure of space have "shapes and variations"? It can, for we now know that space is neither empty nor flat. In the second half of the twentieth century empirical evidence became available, which shows that space is a superdense field of turbulent virtual energies (or, in a more technical formulation, that it is a field of action-quanta that generates energy). It is filled with, and for all intents and purposes *is*, the "quantum vacuum." Thus today we can say, as Clifford, Einstein, and Schrödinger would no doubt say if they were alive, that "material bodies and forces are nothing but shapes and variations in the structure of the quantum vacuum."

We can further agree with Clifford that the "shapes and variations in the vacuum" are waves. There are propagating waves in the vacuum, such as the photons that carry light and the bosons that carry force; and there are standing waves that make up the vast variety of the seemingly solid entities we call matter. Space itself—more exactly, the vacuum that fills space—is not merely a backdrop or container for the motion of matter, but the very "stuff" or substance from which the matter that populates space and time emerged, and through which it continually interacts.

New evidence is forthcoming that the vacuum is a complex, extremely dense, and strongly interacting field. In experiments already noted in chapter 2, physicists at the Relativistic Heavy Ion Collider at Brookhaven National Laboratory discovered the "gluon-field" that binds quarks and endows them with additional mass. They accelerated beams of gold nuclei around a two-and-a-half-mile track and achieved temperatures of 100 billion electronvolts—300 million times higher than the surface of the Sun. (This is believed to have been the temperature of the universe in the first ten milliseconds after the Big Bang.) The scientists found that the gluon-field is extremely dense—thirty to fifty times denser than predicted. The quarks in this field exhibit highly synchronized group behavior and interact strongly with each other and the surrounding gluons. This must have made the early vacuum more like a liquid than a gas. It appears that at the birth of our universe the vacuum was ten to twenty times more liquid than water is today.

It is significant that the vacuum proves to be a dense field with the properties of a liquid even at temperatures 300 million times higher than the surface of the Sun. This suggests that, at the extremely cold background temperature that now reigns in the cosmos, the vacuum field has the properties of a superfluid—a medium in which particles, atoms, and other objects move as fish through water, with subtle and hardly evident effect. Even more importantly, the new finding suggests that when particles, atoms, and other things move in this superfluid medium, their movement makes waves, much as movement in an ordinary fluid.

If the vacuum is the fundamental medium of the cosmos, and if it is superfluid and all things in it produce waves, we should expect that of the two aspects that define particles, namely the corpuscular and the wave aspect, it is the wave aspect that is fundamental. We have reason to believe that this is the case.

An ingenious experiment by Shahriar Afshar, a young Iranian-American physicist, demonstrates that even when the corpuscular aspect of a particle is observed, the wave aspect is still there (since the interference patterns that build up on the screen in the classical "split-beam experiment" do not disappear when the photons—seemingly discrete entities—pass the slits one by one). Bohr's celebrated "complementary principle" does not hold: although both the corpuscular and the wave aspects of particles can be observed, it is not true that they are never present at the same time. The wave aspect *is* present even when the experiment is set up so the corpuscular aspect can be observed, whereas the corpuscular aspect is *not* present when the wave aspect is queried.

Clifford, Einstein, and Schrödinger were right. In the words of former MIT physicist Milo Wolf, the wave medium, that is, the cosmic vacuum in which the waves appear, is the single source of matter and natural law in the universe. Wolf's conclusion: "Since the waves of each particle are intermingled with the waves of other matter and all contribute to the density of the medium, it follows that every charged particle is part of the universe and the universe is part of each charged particle."

Why is it, then, that we see solid material bodies when they are actually waves in the vacuum? The answer is: because vacuum-waves are like "solitons" (so-called solitary waves)—they appear to be discrete and detached bodies, yet they are waves in the medium in which they appear.

The soliton phenomenon was first reported to the British Association for the Advancement of Science in 1845. J. Scott Russell recounted riding beside a narrow channel of water and observing a wave rolling with great speed, "assuming the form of a large solitary elevation, a rounded, smooth, and well defined heap of water, which continued its course along the channel apparently without change of form or diminution of speed." Since then solitons have been observed under diverse conditions. They behave like discrete material objects: they move along defined trajectories, and if they meet each other they deflect one another. Solitons appear not only in water, but as impulses in the nervous system, in complex electrical circuits, in tidal bores, in atmospheric pressure waves, and in heat conduction in solids. They appear in superfluid mediums. Thus they appear in the quantum vacuum, a superfluid cosmic medium.

The solitons that appear in the vacuum are the matter and force particles of the observable universe. When the particles are in their virginal state—not observed and not interfered with—they are in a sense everywhere in the vacuum: they are "distributed," like information in a hologram. But when they are observed and interacted with, they lose their distributed wave aspect and become seemingly discrete material bodies. The particles assume the form of classical objects; quantum physicists say that their wave function collapses.

Although what we perceive with our senses is solid matter moving about in empty space, physical reality is different. In the final count the material universe, including particles, stars, planets, rocks, and living organisms, is not material: all these matter-like things are complex waves in the quantum vacuum.

In light of the latest findings, we can specify the pertinent features of the vacuum, the space-filling medium that is the fundamental element of physical reality. This medium:

—fills all of space and endures through all of time;
—is superdense, filled with fluctuating zero-point energies;
—is superfluid, so that particles and the objects built of particles move through it without evident friction;
—is the originating ground of the particles that make up the manifest universe;
—is the receptacle of the particles when they "evaporate" in black holes.

The cosmic vacuum is responsible for:

—the gravitational attraction among particles and objects built of particles;
—the electrical and magnetic waves associated with charged particles;
—the short-range attractive and repulsive forces in the atomic nucleus;
—and the interfering waves that quasi-instantaneously and enduringly connect particles, atoms, and all things throughout a region of the universe.

Zero-point energies, the G-field, the EM-field, the nuclear fields, and the A-field are specific manifestations of the "unified vacuum." Although Grand Unified and Super-Grand Unified Theories are still in development, we can already affirm that the unified vacuum fills all of space, that it is superdense and superfluid, that it brings forth the particles that furnish local universes and receives them back at the end of their evolutionary cycle, and that it generates the force-fields of gravitation, electromagnetism, and the strong and the weak nuclear interactions, as well as the holographic field that instantly and enduringly interconnects particles and atoms, and all things built of them, in space and time. Science's emerging vision of reality is the vision of a reality that is interconnected and whole—it is an integral vision of reality.

THE NATURE OF SPIRITUAL REALITY

Science's concept of physical reality is not final: like all theories in the empirical sciences, it is subject to revision and improvement. Although this vision is integral, it is not complete, for the natural sciences do not deal with all aspects and dimensions of reality, not even of humanly experienced reality. In addition to the physical aspect or dimension of reality, human experience testifies that reality also has another aspect: a spiritual dimension.

Is the exploration of the spiritual dimension of reality scientific, or is this endeavor mystical, esoteric, religious, or just imaginary? Although mainstream scientists would contest it, the investigation of the spiritual aspect or dimension of reality is also within the scope of science, because—just like reality's physical dimension—it, too, reposes on the testimony of human experience. The experiential evidence for reality's spiritual dimension is not the experience of supernatural entities; the evidence is more modest but also much closer at hand: it is our own consciousness.

Science's concept of physical reality, we have seen, extends beyond immediate sensory experience to include elements that render this experience coherent and consistent. This is true also of explorations of spiritual reality. The difference between science's concept of physical reality and explorations of spiritual reality is not in the conceptual superstructure through which we seek to comprehend the world, but in the starting point. Science's concept of physical reality takes off from the content and reference of sensory perception; it takes the world we perceive as a physically real domain situated beyond our perception of it. Explorations of spiritual reality, on the other hand, take off not from the content and reference of perception, but from the very *fact* of perception. We take off from the givenness of conscious experience—in one word, from *consciousness.*

Can we derive a concept of the nature of spiritual reality from the fact that we have consciousness? Consciousness, after all, is "private": it is "my" consciousness; it is only experienced by me. If we take this restriction seriously, the whole world becomes an element of "my" consciousness. But this position, while logical, is not satisfactory to everyone.* It is reasonable to hold that science's concept of physical reality is too important to be dismissed as a projection of "my" consciousness. If so, we should view physical reality as an aspect or dimension of the real world, the same as spiritual reality. *They are two aspects or dimensions of one and the same reality.*

On this assumption we get a logical and coherent account of the nature of spiritual reality. Other things exist in the world beside my own consciousness, and these other things also possess consciousness, although they could possess consciousness of a more rudimentary (or perhaps a more advanced) form. In an impartial view reality is not just physical: it is *psychophysical.*

What is the origin of the consciousness that resides in some form in all things in the psychophysical cosmos? If the spiritual dimension of reality is a dimension of the same reality as the physical dimension, the answer to this question is evident. Consciousness has the same origin as the things that make up physical reality: its origin lies in the quantum

*The position that the entire world can be considered an element of conscious experience is eloquently argued by physicist-philosopher Peter Russell in part three, below.

vacuum. The vacuum is not only the originating ground and ultimate destination of the particles, atoms, and the complex entities that, consisting of ensembles of particles and atoms, make up the physical reality of the universe; it is also the originating ground of the consciousness associated with particles and systems of particles throughout the cosmos —of the universal consciousness that is the fundamental feature of the spiritual dimension of reality.

Do we have evidence for this claim? Can we tell that the vacuum is not only a superdense virtual energy field from which spring the wave-packets that appear as matter, but is also the seat of the consciousness that infuses my body and brain the same as the rest of the universe?

Clearly, there is no way we could tell by reference to everyday sensory experience, or even by observations and experiments based on direct experience. Consciousness cannot ordinarily be observed in anyone but ourselves. But there is a more speculative yet entirely meaningful approach we can take: we can envisage a thought experiment. It is based on the finding of transpersonal psychologists that in altered states of consciousness we can enter the consciousness of other people, even of animals and nonliving things. According to Grof, in altered states of consciousness we can experience fusion and identification with practically any aspect of the world.

The thought experiment we envisage is to enter a profoundly altered state of consciousness and attempt to identify ourselves with the quantum vacuum. Assuming that we succeed, would we experience a physical field of fluctuating energies? Or would we experience something like a cosmic field of consciousness?

We cannot exclude the latter possibility. Grof and other transpersonal consciousness researchers claim that in deeply altered states people experience a form of consciousness that appears to be that of the universe itself. This most remarkable of altered-state experiences surfaces in individuals who are committed to the quest of apprehending the ultimate grounds of existence. When the seekers come close to attaining their goal, their descriptions of what they regard as the supreme principle of existence are strikingly similar. They describe what they experience as an immense and unfathomable field of consciousness endowed with infinite intelligence and creative power. The field of cosmic consciousness they experience is a cosmic emptiness—a void. Yet, paradoxically, it is also an essential fullness. Although it does not feature anything in a concretely manifest form, it contains all of existence in potential. The void they experience is a fullness; the vacuum is a plenum. It is the ultimate source of existence, the cradle of all being. It is pregnant with the possibility of everything there is. The phenomenal world is its creation: the realization and concretization of its inherent potential.

People who practice yoga and other forms of deep meditation report the same kind of experience. This was the basis in the Indian Vedic tradition for the affirmation that consciousness is not an emergent property that comes into existence through material structures such as the brain and the nervous system, but a vast field that constitutes the primary reality of the universe. In meditation, when the gross layers of the mind are stripped away, this field is experienced as unbounded and undivided by objects and individual experiences. Underlying the diversified and localized gross layers of ordinary consciousness there is a unified, nonlocalized, and subtle layer: "pure consciousness."

There is a noteworthy parallel here between insights that have been present to the human mind for thousands of years and the implications of the latest findings of the sciences. In regard to the physical dimension of reality, the coincidence concerns an element of ancient Hindu philosophy. In his already cited *Raja Yoga,* Swami Vivekananda made clear that the Hindu seers considered Akasha the basic reality of the cosmos. *"When there was neither aught nor naught, when darkness was covering darkness, what existed then?" asked Vivekananda, and answered, "Then Akasha existed without motion. . . . At the end of a cycle the energies now displayed in the universe quieted down and became potential. At the beginning of the next cycle they start up, strike upon the Akasha, and out of the Akasha evolve these various forms . . ."*

The latest cosmologies re-discover the cyclically self-renewing universe, a cosmos that takes off from, and returns to, an enduring fundamental medium. The ancient Hindu cosmology can be restated in contemporary scientific terms simply by substituting "quantum vacuum" for "Akasha." "*When there was neither aught nor naught, when darkness was covering darkness, what existed then? Then the quantum vacuum existed without [manifest] motion. . . . At the end of a cycle the energies now displayed in the universe quieted down and became potential. At the beginning of the next cycle they start up, strike upon the vacuum, and out of the vacuum evolve these various forms . . .*"

As we move from the physical dimension of reality to its spiritual dimension, the parallel between ancient insight and science's new vision of reality continues to hold true. According to many traditional cosmologies, in the course of time the universe's undifferentiated, all-encompassing consciousness separates off from its primordial unity and becomes localized in particular structures of matter. In the new scientific context we can say that the proto-consciousness of the vacuum becomes localized and articulated as the particles that emerge in it join together in atoms and molecules. On life-bearing planets the cosmic proto-consciousness becomes further articulated as complex molecules form living cells, and then single- and multi-cellular organisms. It achieves the articulation of human consciousness when it is associated with the supercomplex brain of the human organism.

Human consciousness is a high-level articulation of the consciousness that stems from, and is rooted in, the quantum vacuum: the superfluid universal field that is the fundamental element of the integral reality of the psychophysical cosmos.

Part Three

THE RE-UNION OF SCIENCE AND SPIRITUALITY

A Roundtable of Leading Thinkers

There is a popular preconception in contemporary society: science is "nothing but" observation and the measurement and computation of observations. In turn, spirituality is "nothing but" either devoutly following a religious scripture or sitting at the feet of a guru. How could there be any meeting ground between them? How could a white-coated scientist in a laboratory find something in common with a person reciting ardent prayers in a holy place, or with a long-haired escapee from mainstream society?

The popular preconception is mistaken. As the foregoing parts of this book have shown, and as the papers in the roundtable that makes up this part confirm, science is far from being a mere exercise in recording and computing observations. Science is part of the perennial human quest for making sense of the world, and of ourselves in the world. It is a search for meaning just as theology, art, literature, and ordinary common sense is—and spirituality as well. The difference between them is not in the end they seek, but in the way they seek it. Science uses rational thinking to analyze and interpret what experience and experiment disclose, while theology combines such thinking with an element of unquestioning faith. Genuine spirituality, in turn, combines experience with the immediacy of an intuition that speaks to a reality that is different from the world conveyed by the senses—it is higher, or deeper.

A first meeting-ground between science and spirituality is, and has always been, their shared search for meaning. This was not recognized by the majority of spiritually inclined people who knew little and cared less about science. On the other hand it was clearly perceived by scores of humanists, philosophers, and even philosophically inclined scientists. The Hellenic thinkers, precursors of the modern scientist-philosophers, were as at home with rational discourse about the world as with its deeper reality; they spoke of the "primary substance" of the universe just as readily as of beauty, the good, and the soul. There was no gap in their thinking between everyday observation and deep intuition, no divide between "science" and "spirituality."

The gap between spirituality and science arose when modern science was born: scientific inquiry had legitimacy in the eyes of the Catholic Church only if it limited itself to "natural philosophy," leaving "moral philosophy"—all things to do with values and ethics, and mind and soul—to theology. Giordano Bruno was burned at the stake for transgressing this divide, while Galileo managed to escape. He made a careful distinction between "primary qualities" such as the solidity of bodies, their extension, figure, number, and motion, and secondary qualities such as color, taste, beauty, or ugliness. He claimed freedom for science only for the investigation of the former. This distinction was reinforced when Descartes separated "thinking substance" *(res cogitans)* from "extended substance" *(res extensa)* and identified the one with human consciousness and the other with everything else in the world. Clearly, a thinking substance capable of perceiving secondary qualities can be spiritual, but there is no way that an extended substance that only has solidity, extension, number, and motion can.

The vexing divide between the material world and the world of mind and spirit did not prevent great scientists from seeking a meeting ground between them. The founders of modern science were integral thinkers. Bruno, Galileo, Copernicus, Kepler, and Newton himself, had deep spiritual, even mystical streaks. Nor did spirituality lack in the giants of twentieth-century science. As their writings testify, it was present in Einstein, Schrödinger, and Bohr, and in Pauli and Jung, to mention but a few.

Despite the life-long search of many scientists-philosophers for a union between the experiences and experiments on which science is based and the intuitions that are the basis of spirituality, a meeting ground between the two was hard to find. Until the last ten to fifteen years, that is. Until science began to discover a cosmos that is not the domain of bits of unconscious matter moving about in passive and empty space. Until it began to recognize that the universe is a dynamic, quasi-living self-evolving system, interconnected and integral at all scales and in all domains. Until it began to recognize that this system conserves and conveys not only energy, but also information. And until the recognition dawned that consciousness is just as fundamental in the universe as energy, and more fundamental than matter . . .

A consciousness- and information-imbued, interlinked, and integrally evolving cosmos is a reenchanted cosmos. It provides a ground not just for an encounter and dialogue between science and spirituality, but—as the eminent thinkers that make up the roundtable that follows make clear—for their *re*-union.

The Furrows of Reality

Scientific and Spiritual Implications of the Reenchanted Cosmos

Stanley Krippner and Brian A. Conti

In 2004, the film *What the Bleep Do We Know?* was released. It was a "sleeper" (a surprise box-office hit). The film consisted of interviews with experts on quantum physics, as well as a storyline involving a young woman caught up in the frenetic pace of modern life. The insights of quantum physics were used to assist the young woman (played by the Academy Award-winning actress, Marlee Matlin) navigate her way through life. Ervin Laszlo's book, *Science and the Reenchantment of the Cosmos,* has similar potential to improve people's lives, because its message is one that affirms the unity of human beings, rather than the divisions that separate them.

The ancient concept of unity of consciousness or Oneness—which finds expression in established Eastern thought, such as in the literature on the Tao and on Brahman—is reborn and restated in the framework of Laszlo's Akashic Field (or the A-field). People of various faiths have become aware of the Akashic Field through personal encounters with it, notably in the form of personal or group spiritual, ecstatic, and mystical experiences. Through the ages, certain people of all religious persuasions have been more inclined to look inward rather than outward for answers, and many have come to remarkably similar conclusions, principally an idea of Oneness. In Laszlo's formulation of the A-field, however, this unity can be apprehended through the eyes of contemporary science rather than solely through spiritual, ecstatic, and mystical experiences, making it more readily assimilated by Westerners for whom "unity" may be less than familiar.

In ancient India, the Sanskrit word *akasha* or "cosmic sky" resembled the contemporary concept of "space." The term referred not only to intergalactic space, but also to space (and time) in the loftier dimensions of life. It encompassed the wellspring of creation and the domain of the divine, and signified a space-time continuum that was all-pervasive.

Akashic Records (of all that happens, and has ever happened) have been written about over the millennia and were the content of oral mythology even earlier. Robert Cheney, in his 1996 book, *Akashic Records,* suggested: "Perhaps the vibrational threads of a single life . . . permeate the entire universe, interacting with other vibrational energies, both here and as distinct as your consciousness will accept. . . . Just as the words and music of a song are preserved on audiotape, or a compact disc, so the story of your life is saved in the Akashic Records."

Certain biblical passages are reminiscent of the Akashic Records: the book of Joshua describes how the prophet took a stone and proclaimed, "Behold, this stone shall be a witness to us, for it has heard all the words of the Lord that he spoke unto us." The book of Daniel cites the Archangel Michael as saying that all people "will be delivered" if their names "shall be found written in the book." The book of Malachi speaks of a "book of remembrance," and St. Paul's Second Epistle to the Corinthians describes a book "written not with ink, but with the Spirit of the living God." Hence, the notion of a universal record is not limited to Hindu traditions. Indeed, the Islamic Holy Koran refers to such a record in terms that Cheney suggests are reminiscent of modern computer files.

Cheney makes the case that the Akashic Records "are the imprint of the Self on the eternal, universal, electromagnetic atmosphere of primary substance" and that this "substance is everywhere." He continues by saying that "each entry in the field is known as a *field.*" Thus, many Eastern and Western religions seem to share a common thread: that of a universal record, system, ground, or field. Ervin Laszlo's books provide a contemporary perspective on this ancient concept.

The notion that everything is somehow subtly connected to everything else is a very old one indeed. However, it is also a relatively new one from the perspective of contemporary science. This is essentially what is proposed by the concept of the A-field, joined with the idea that consciousness is the ground of all being. To paraphrase the physicist Sir James Jeans, it often seems that the universe is structured more like a great mind than a physical realm. This view implies that matter is more like a thought than an inert lifeless substance, and that even a rock possesses consciousness of a sort.

This ancient way of looking at life and the universe also seems to resolve the paradoxes introduced by modern quantum physics. It seems that science is finally knocking on the door of spirituality, an exciting idea indeed.

In this and in previous books Laszlo explores specific puzzles and paradoxes facing today's world, and what they indicate about the hidden nature of reality. He makes a plausible case for a ubiquitous underlying informational field through which everything is essentially connected. He also explores the inevitable Western resistance to such an idea, as it countervails many long-held concepts that are central to Western thought. Laszlo effectively anticipates much of this resistance by using historical evidence to demonstrate that most major advances in understanding have initially been met with reluctance or scorn.

The A-field, indeed, bears all of the characteristics of a radical intellectual advance. Not only can Laszlo's concept of the A-field explain nuances in reality, but it also presents a tenable synthesis of spiritual thought and scientific data. Laszlo contends that science will eventually experience a paradigm shift and accept the idea that all creation is informed, interconnected, and of an informational rather than a material nature. Once these concepts are generally accepted, humanity can adjust to a more satisfying and comprehensive view of reality, and very likely also to a more satisfying mode of existence.

As the entailments of an all-pervasive information field are explored, many current puzzles will resolve themselves, not only within science, but in other human realms as well. These might include the mind-body problem, the efficacy of placebos in healing, the purported effectiveness of intercessionary prayer, various forms of psychic phenomena including hauntings and poltergeists, synchronicity, so-called past-life experiences—as we have shown in the series of studies published in *Varieties of Anomalous Experience,* the list is almost endless. Implicit in the idea of an informed unity is the possibility that human beings, as stores of conscious potentiality within the universal continuum of consciousness, can actually contribute to physical manifestations. In this view people might actually be unknowing participants in creation, without fully understanding what they are doing.

If we human beings do live in a consensual reality that conforms to our unconscious expectations and materially unfolds according to our notions of what is logical and possible, we will eventually face the fact that we are the architects of the logical, mathematical, and linguistic "laws" that structure our experience of reality. The A-field is all-pervasive and quasi-infinite; hence there is a limit to what extent it can be canalized and molded. Paradoxically, the underlying unity of the A-field has been socially partitioned by hundreds of languages, mathematical systems, linguistic conventions, social mores, religious institutions, forms of government, and classifications of people, their behaviors, and their environment.

As people construct belief systems they have no way of verifying whether the principles of their belief accurately explain reality or artificially structure chaos. Even the arrow of time itself may be a synthetic imposition on an underlying timeless reality; the collective experience of linear time may betray the true nature of the universe.

Another appendage of the A-field is that any strongly held conviction might manufacture its own ends. For example, extremely pious people sometimes have ecstatic experiences, or perhaps see "signs" that their God is at hand, that they are on the right path, or that everyone else is clearly misguided or heretical. However, they often fail to recognize that the conformity of their reality is the product of their belief. The outcome corresponds with their strength of belief, their social support system, and the time and place in which they find themselves. This underlines the potentially negative impacts of dogma in religion and spirituality. The veracity of belief may be "revealed" or proven "true," not through its alignment with reality, but through events that the belief anticipates or creates.

Human beings are the only organisms on Earth who perceive the inevitability of their own death; they have constructed numerous personal and cultural myths and belief systems to "buffer" themselves against this knowledge. They may cling tightly to their devotion, not realizing that their beliefs, and those that they oppose, are merely manifestations of the underlying A-field. The twenty-first century already is beset with three dozen religious, ethnic, and political wars. Many, if not most, of these conflicts could be mollified if the opponents recognized their common humanity and decided to live, play, and work, if not together as friends and colleagues, then as neighbors and acquaintances.

Any serious investigation of the A-field and its implications requires a particularly balanced approach. Science, despite its faults and flaws over the ages has, at the very least, earned a reputation as the system of inquiry least likely to inspire delusion and self-fulfilling outcomes. Of course, scientists are vulnerable to these same errors, and if intransigence leads to outdated beliefs, their perceptions will be skewed by the very same feedback mechanisms as those that have bedeviled spiritual and religious belief systems over the ages. It took hundreds of years for Western astronomers to abandon the notion of a heliocentric universe, even as additions such as "cycles" and "epicycles" abounded to privilege Earth's central position. Once scientists abandoned the geocentric Ptolemaic model, they encountered opposition from religious leaders who were loath to surrender Earth's exalted status as the center of the universe.

Some day scientists may scale the peaks of human understanding and they will be welcomed by groups of mystics and seers that have been waiting there all along. The very fact that contemporary science is investigating areas that seem to border on the spiritual, and that quantum physics and other cutting-edge sciences are resolving puzzles and paradoxes, indicates the arrival of a time in which both those who are spiritually inclined and those who are scientifically disposed can appreciate the concept of the A-field. That is not to say that these disciplines need to be merged; science asks different questions than religion. Science asks "What?" "When?" and "How?" Religion, philosophy, and spiritual systems ask "Why?" and "What for?" The former addresses questions of function and form, while the latter address questions of intention and purpose.

That Ervin Laszlo has written this book is a long overdue step toward a fuller realization of human potential, the expansion of knowledge, and the awareness of interconnectedness. Many traditional shamans, adepts engaging in contemplative practices, and mystics caught up in ecstatic reveries have known these aspects of reality all along, but more empirically-minded men and women needed a presentation that befits their instruments, equations, and reason. With concepts such as the A-field, the meeting between science and spirituality is finally coming to pass.*

---

*The preparation of this essay was supported by the Saybrook Graduate School and Research Center Chair for the Study of Consciousness.

From a Mechanistic and Competitive to a Reenchanted and Co-Evolving Cosmos

Elisabet Sahtouris

In the introduction to this book, Ervin Laszlo says:

> The reenchantment of the cosmos as a coherent, integral whole comes from the latest discoveries in the natural sciences, but the basic concept itself is not new; indeed, it is as old as civilization.

This book is vital to our transition from a long epoch of empire building—of the drive to control Earth's resources by fierce competition in a situation of perceived scarcity—to a future of truly cooperative global family. Laszlo's universe is a far cry from the one Western science has taught us, and compatible with my own views as a "post-Darwinian" evolution biologist. In fact, no small number of Western scientists today have defected from the official scientific story of How Things Are to create or adopt various versions of the new scientific worldview Laszlo presents—a worldview rooted in older sciences and cultures such as Vedic, yet ultimately not just timely, but futuristic. These new scientific models, integrating ancient and modern knowledge, represent a paradigm shift greater than the Copernican revolution and are crucial to the next stage in humanity's evolution as a planetary species.

How Nature Lost Its Life. In the eighteenth and nineteenth centuries, a new alliance of bourgeois entrepreneurs and scientists forged the industrial revolution, bringing the modern age successfully into being and replacing the prior cultural hegemony of the alliance between Church and State. Eventually science became far more than a vast research enterprise that gave us an advanced technological society with more commercial products than any previous culture could possibly have imagined, along with "progress" at a breakneck pace that leaves us breathless and wondering if we can even hope to catch up with our own children and grandchildren. Western scientists, in addition to their role in spawning that technological society, became the new priesthood mandated to give us our cultural worldview: our beliefs about How Things Are in this great universe of ours, and on our planet Earth in particular. This is a relatively new and very important historical phenomenon in the history of civilization, as the priesthoods of most previous civilizations—defined as large organized sociopolitical entities with urban centers—were religious (with notable exceptions such as China), getting their worldviews more from revelation than from research.

The Western scientific worldview founded by Galileo, Descartes, Newton, Bacon, and others was—in stark contrast to previous views of living Nature—one of a non-living, non-intelligent mechanical universe, a clockwork projected from human mechanical inventions to the Cartesian "Grand Engineer." In God's design of Nature humans were just complex robots, the males alone imbued with a piece of God-mind, according to Descartes, so that they, too, could invent machinery.

As models of celestial mechanics (the Newtonian motion of stars and planets) became more elaborate, social institutions as well were increasingly seen and modeled on mechanisms and expected to run like the well-oiled machines of factories. Time/motion efficiency studies of workers turned people themselves into machines, as Charlie Chaplin movies so well caricatured. Most of today's businesses, governments, educational institutions, and so on are still conceived, organized, and run as hierarchical mechanics.

As men of science came to feel increasingly competent and knowledgeable about the physical world—and in consequence felt themselves to be in control of human destiny—they overthrew Descartes' Grand Engineer, formally abandoning the "hypothesis of God," and thereby removing any notion of Nature, including humans, as existing through sacred creation. Rather, Nature was redefined as a wealth of natural resources to be exploited by Man, the pinnacle of accidental, natural evolution. Oddly, the concept of machinery—which by definition exists *only* through purposive invention and assembly—was retained by scientists as the appropriate metaphor for Nature. But they believed that machinery was assembled through a series of blind accidents.

One of the most pervasive and persistent cultural beliefs we have been given by science, as Laszlo has pointed out, is thus the concept of a non-living, accidental, purposeless universe running down by entropy, with life defined as transient "negentropy" opposing this force of decay, yet never overcoming or even balancing its inevitable slide into heat death. To me, this is like describing the life of any one of us as a one-way process of decay toward death, with a *negdecay* process of birth and growth opposing it, though overall unsuccessfully.

This dreary view of life, which spawned philosophies of meaninglessness such as Existentialism and Dadaism, made me wonder deeply about the very concept of non-life invented by Western science. All previous cultures understood life and death, but non-life is a concept pertaining to something that never was alive—a concept that came into human culture with the invention of mechanism in ancient Greece and resurfaced in the new era of European mechanics. Was it really appropriate, I asked myself, for science to force life to be defined within a context of non-life? Could one really explain the existence of living things as accidentally derived from non-living matter? Could one derive intelligence from non-intelligence, consciousness from non-consciousness, as I was consistently taught in the graduate science departments of several universities and research institutions?

To answer these questions I was led to ponder what might have happened had Galileo looked down through a microscope into a drop of pond water teeming with gyrating life forms instead of up through a telescope into the heavens, already conceived in his time as celestial mechanics? Might biology, rather than physics, have become the leading science into whose models all others must fit themselves? Might scientists then have seen life not as a rare accidental occurrence in futile struggle to build up syntropic systems against the inevitably destructive tide of entropy, but as the fundamental nature of an exuberantly creative universe?

Instead of projecting a universe of mechanism without inventor, assembling blindly through particular, atomic, and molecular collisions, a few of which came magically to life and further evolved by accidental mutations, I propose that there is reason to see the whole universe as alive by the definition of *autopoiesis:* a living entity being one that is in continual self-creation in relation to its surround. In self-organizing endless levels of size-scale and complexity, the universe or cosmos learns to play with possibilities in the intelligent co-creation of its evolving living systems.

It seems to me more reasonable to project our life onto the entire universe—yes, the heresy of anthropomorphism—than to project onto it *mechanomorphism,* our non-living machinery, the product of *anthropos* and therefore a truly *emerging* phenomenon rather than a fundamental one. I propose that it is possible to create a scientific model of a living universe, as I have done elsewhere, and that such a model is not only scientifically justified but can lead to the wisdom required to build a better human life on and for our planet Earth, as the ancient Greeks intuited it should.

The original intention of science—called *Philos Sophias,* or Lover of Wisdom, by the ancient Greeks—was to understand the Cosmos (the patterned universe) in order to find guidance in human affairs. As an evolution biologist and Greek philosopher, I returned to these roots in which scientific research and epistemology were integrated in a single endeavor, to pursue the age-old quest for meaning: who we humans are, where we came from, and where we are headed, as well as my personal questions about where we are now in the human species' particular evolutionary trajectory and how we can move ourselves along it consciously and positively.

As Ervin Laszlo tells us in his conclusion:

> Science . . . is not only a collection of abstract formulas, and not just a wellspring of technology; it is a source of insight into *what* there is in the world, and *how* things are in the world. By this token science is part of the perennial human quest for meaning and understanding.

Entropy and Evolution Reconsidered. In the two decades from 1850 to 1871, Rudolph Clausius reformulated the fundamental laws of the universe as energy constancy and one-way entropy, while Charles Darwin published his seminal works *On the Origin of Species* and *The Descent of Man.* Between them they gave us a non-living universe running down to heat death by entropy, with life as a temporary "negentropic" struggle.

I will argue that Clausius' model of a universe running down by entropy and the Darwinian model of biological evolution as an endless competitive struggle for scarce resources both give us half-truths about Nature that seemed appropriate in their historical context but are now seen to be fundamentally flawed, thereby seriously misleading us and holding up our own natural evolution. The full truth—including the other half of a more holistic view in physics and biology respectively—reveals that Nature is on our side in role-modeling the evolutionary leap that would rapidly bring about an energy-efficient and globally beneficial human economy that functions like a truly healthy living system.

Clausius' work on the thermodynamics of entropy, openly acknowledged by Maxwell in England, was based on Sadi Carnot's experimental work with energy transfer in the closed mechanical systems of steam engines and applied (by Clausius) to the universe as a whole with no evidence that the universe was a closed system in which such extrapolation might be valid. Yet these two "inviolable laws," along with the more basic conceptualization of the universe as purposeless non-life, have persisted since then as absolute dogma in physics and all other areas of science.

The entropic universe, conceptualized as beginning with a Big Bang and deteriorating ever since, is in sharp contrast to previous worldviews of Nature as alive and vibrant with intelligent creation and purposive direction. This earlier view is compatible with my own model of a self-organizing, living universe in which planetary life is a special case of extra complexity, now actually measurable as being halfway between the microcosm and the macrocosm, where "upwardly" and "downwardly" spiraling energies collide on physical surfaces where such life can evolve.

Historically, the social consequences of the proclamation of an entropic universe by the scientific establishment were enormous, especially when coupled with Darwin's theory of life's evolution.

Darwin himself had concluded, with great elaboration in his magnificent opus on *The Descent of Man,* that humans must exercise their evolved capacity for moral behavior, as David Loye has effectively shown in his book *The Great Adventure*. However, this aspect of Darwin's work was not promoted by the science that took up his theory of evolution, which focused instead on his explanation of struggle in scarcity as the driver of evolution. I see this as a half-truth rooted more in Darwin's historical context than in Nature itself. Had Darwin been able to see beyond that context, he might have noticed that many systems, evolving long before humans, had evolved out of that struggle into cooperation, mutual support, altruism, and other features we define as ethical.

In *The Origin of Species* Darwin wrote:

> Nothing is easier than to admit the truth of the universal struggle for life, or more difficult . . . than constantly to bear this conclusion in mind. Yet unless it be thoroughly engrained in the mind, I am convinced that the whole economy of Nature, with every fact on distribution, rarity, abundance, extinction, and variation, will be dimly seen or quite misunderstood . . . As more individuals are produced than can possibly survive, there must in every case be a struggle for existence . . . It is the doctrine of Malthus applied with manifold force to the whole animal and vegetable kingdoms; for in this case there can be no artificial increase of food, and no prudential restraint from marriage.

Darwinian theory as Darwin *himself* established it—not just through later misuse as "social Darwinism"—was thus very essentially rooted in political economy. As Darwinian evolution theory became known, along with Clausius' entropy, Western culture established its view of the purposeless and hopeless human struggle against natural odds in a blindly deteriorating and non-caring universe. The concept of getting what you can while you can against fierce competitors for scarce resources extended into philosophy, psychology, art, and Western culture at large.

Such beliefs became the driving force of modern and post-modern Western consumer culture. Humanitarian social values and morals were left to religions with lesser persuasive clout than science, which came to openly pride itself on being value-free, though it was never an independent cultural endeavor free to pursue unbiased inquiry into Nature. Rather, as we have seen, science was empowered to tell us how things are in our universe and who we are within it by an ever more powerful political economy, which in turn was kept in power by the valuable engineering applications of science. Small wonder that economics was manifested as a competitive struggle justified as "social Darwinism" and deemed inescapable.

Science and the New View of Reality. The whole edifice of Western science was built on belief in an objective and measurable world, separate from humans, which could be studied as a reality independent of human experience and action. This belief in objectivity has been discredited, first logically by philosophers of science and increasingly by scientific research itself. If the claim of basing science on reason—on experiment (a word derived from *experience*) and rational argument—is to be upheld, then we cannot postulate a world that is *not* within human experience for the simple reason that we have no way to be outside human experience.

The seemingly obvious way out of this dilemma is simply to conceive reality *as* our human experience of the world, as many indigenous cultures have done. What other legitimate basis for creating scientific models of our world/cosmos is there?

Somewhat surprisingly, this idea is not as strange to our own culture as I had thought. *Webster's Online Dictionary* actually defines reality as "something that is neither derivative nor dependent but exists necessarily."

The only thing fitting this definition is direct experience, for once any experience is reported to another, whether by a three-year-old, a scientist, or a theologian, it clearly becomes derivative.

The bottom line of human experience is that it all takes place within our consciousness and that our minds form the beliefs on which we act, collectively creating a uniquely human world. Change those beliefs and that world changes accordingly. Our world is now in sufficient crisis that transparency in all our endeavors is critical to our survival. Light shed on the relationship between science and political economy can, I believe, show us the way to true freedom and a healthy economy for all the world's people.

My holistic view of life, like Laszlo's, is of a unified living cosmos that provides a framework for understanding, and more consciously creating, our own human trajectory within its greater process. Biology, physics, and consciousness are some of the lenses through which scientists see our human experience of this universe.

The axiomatic foundation of my model is formed by consciousness, as the inescapable field of human experience, and human reality—the only kind of reality we can reasonably model—as the total of all human experience. This foundation replaces the fundamental assumption of an objective non-living universe in which life happens by accident. Life is defined as an intelligent, evolutionary, self-organization process, rather than as a collection of biological entities evolving by serial accidents in a non-living universe. Thus life is seen as the natural self-creating and recycling metabolism or process of the cosmos itself.

This provides a framework for proposing a new Integral Science with a self-consistent, evolving model of life as self-organizing expressions of the cosmic field of consciousness, which Laszlo calls the Akashic Field, Bohm the Implicate Order, and other physicists the ZPE or ZPF (zero-point energy field). Autopoiesis—the definition of living entities as those that continually create and maintain themselves—can be extended to the self-creation of a living and co-evolving reenchanted cosmos. This gives us an integral view of the whole of reality, with essentially the same generating and recycling processes at all its levels.

The "Metaverse Story"

Where Science Meets Spirit

Christian de Quincey

We are confronted every moment by an enormous mystery: The very existence of the world. German philosopher Martin Heidegger pinpointed the mystery when he posed what is probably philosophy's most profound question: "Why is there something rather than nothing?"

According to many of the world's great spiritual traditions the world of "things" emerged from a prior realm of nothingness or Emptiness. The world of rocks, oceans, mountains, trees, fish, frogs, lizards, birds, dogs, apes, and people, and each one of us who observes and participates in the grand spectacle—all this, came from Emptiness, from nothingness. That is, indeed, *the* great mystery.

How did it all begin? Where are we going? And how? In the beginning, so the story goes, there was no beginning. Then, from nowhere and nowhen, the material universe exploded into creation in what modern cosmologists refer to as the Big Bang. From "nothingness," the universe was born. But because the emergence of something from *true nothing* is logically absurd, we need to revise our understanding of "nothingness" to refer to what spiritual traditions call the "plenum void," and modern science calls the zero-point energy field. Merging these concepts, Laszlo offers us the notion of the quasi-eternal, or possibly eternal, Metaverse.

For Laszlo, "Metaverse" refers to the source of the manifest universe, and it includes information and consciousness as well as energy (as we will see below). The Metaverse gives rise to the entire spectrum of reality: the world of physical entities (physiosphere); the world of living systems (biosphere); the world of information (infosphere); as well as the "noosphere," the world of mind and imaginative creations; and, perhaps, the "theosphere," the world of souls and spirits.

Evolution: Something New Emerges. The entire Metaverse is engaged in a perpetual dance of involution and evolution. It begins from no-thingness, the domain of pure potential (perhaps equivalent to what some traditions refer to as "spirit"). The Metaverse then involves itself in its own creative play by bringing into being a recurring cycle of universes from big bangs to crashing crunches.

But because the Metaverse is beyond space and time (which do not exist until a big bang), its generative power of "emanation and return" occurs outside time; it is a logical, rather than a chronological sequence—we could call it "onvolution" or the unfolding of being. And since the Metaverse is timeless, occurring at every timeless instant, the entire "process" of "efflux" and "reflux" (as Plotinus called it) is happening right now. The entire plot of the Metaverse story is available in every moment.

The universe story is another matter. It unfolds in time, and is truly an evolution—from the Latin *e* (out) and *volvere* (to roll). Our world, the physical universe in which we find ourselves, entered the cosmic story at the moment of the Big Bang, when time began. It all began in that primordial fireball, and the wonder is that despite what must have been an extremely chaotic beginning, the matter of the universe somehow managed to organize itself into ever more complex structures from photons to elementary particles, to atoms, molecules, and interstellar gases, planets, stars, galaxies, clusters, and superclusters. Laszlo's hypothesis of the Metaverse as a storehouse of information-templates from prior universes goes a long way toward explaining this mysterious emergence of order and structure from the apparent initial chaos of the Big Bang.

Not only that; after some four billion years or so, some of this complex matter developed the ability to reproduce itself, and gave birth to the world of biological systems—the biosphere. And even that "miracle" was far from the end of the story. Some members of the living world evolved further and developed the ability to represent themselves and their world through the use of signs and symbols—and so the world of mind and culture arose—the noosphere.

Even more intriguing: During the past couple of centuries, some individuals took the noosphere a step further and discovered the very process of evolution that had brought them and us and the entire world forth from the original moment of creation. In other words, evolution became conscious of itself. And with that, the entire process moved closer to the possibility of directing and choosing its evolution.

Seven Stages of Evolution. We could summarize the stages of this "greatest story ever told" as follows:

- *First:* The plenum void of the unmanifest Metaverse.
- *Second:* The Metaverse instantaneously comes into physical manifestation in the Big Bang, creating time, space, and light. At this cosmic flaring forth, the randomness and chaos of this universe carries within it the "seeds" or templates of information from prior universes.
- *Third:* Using the information-templates, the new universe self-organizes amidst the surrounding chaos. From initial massless, chargeless photons, this universe begins to create its elementary particles, the building blocks of matter and of all the material forms to come later.
- *Fourth:* Matter eventually develops the ability to reproduce itself, and life is born.
- *Fifth:* Some life forms further develop the ability to represent themselves and other objects symbolically—thus mind and language are born.
- *Sixth:* Just moments ago in cosmic time, our species became aware of the process of evolution—and at that moment evolution became reflexively aware of itself.
- *Seventh:* Being self-aware, evolution can now choose its evolutionary path and goal, although the options available remain constrained within the inevitable cyclical birth-and-death of the universe. In

> time, the universe, including all its conscious creations, will return to its source, falling back into the womb of the Metaverse. But all is not lost. The information and energy patterns generated within the lifespan of the universe will be stored in the A-field of the Metaverse, ready to in-form a next, new universe—and so, the process will begin all over again.

It all sounds so marvelously simple despite the awesome time spans, spatial immensities, and unimaginable complexities involved. The details may be complex and bewildering, but the grand sweep of evolution—the plot of the Metaverse story—is straightforward enough. Seven great stages from the source to return.

Bridging Science and Spirituality. Laszlo's grand vision is appealing for a number of reasons. Not only does it offer a unified view of quantum physics, cosmology, evolution, and even the role of information and consciousness—it also presents an all-encompassing scientific model that seems to validate perennial wisdom from some of the world's great spiritual traditions. But why should we get excited by discoveries in science that appear to validate profound insights from perennial philosophy? What prompts us to turn to science for confirmation of spiritual wisdom?

In this and his previous book, *Science and the Akashic Field,* Ervin Laszlo makes the case that science is finally in a position to produce a theory of everything (ToE). Drawing on anomalies and advances in cosmology, quantum physics, biology, and consciousness studies, he shows how the discovery in physics of the zero-point energy field (ZPE) is also the discovery of a universal *information* field.

Quantum physics, the "gold standard" of science, has discovered that underlying the manifest world of matter and energy is a universal field of quantum potential—the ZPE field. It is the source or foundation for all of physical reality. Everything that exists in our universe comes from ZPE.

The concept and implications of zero-point energy stretch our minds and imaginations to new limits. Like so much else in quantum physics, ZPE nudges us to open up to a new paradigm, to accept a radically new, highly counter-intuitive understanding of the deep nature of reality. Yes, ZPE is bizarre—nevertheless it is rock-steady, solid science.

It explains how the world we know and live in—our undeniable, familiar reality—springs forth at a dizzying rate, billions of times a second, from the universal field of quantum potential. Everything we know, everything that exists, comes from the ZPE field, and sooner or later returns there, to be "recycled" back into our world in some other form, or perhaps into another universe.

The discovery of this field may be final confirmation from science of a profound insight into the nature of reality from ancient Hindu cosmology. The Akashic Record, like the ZPE field, records everything that has ever happened, is happening, and will happen from the birth of our cosmos till its ultimate end. In scientific terms, the "everything" that is recorded is the sum total of all events and the information they contain.

Based on the remarkable similarities between the Akashic Record and the ZPE field, Laszlo equates the two, and renames them collectively the "Akashic Field" or "A-field"—deliberately using a scientific-sounding term for a venerable spiritual-metaphysical idea.

Science Confirming Spiritual Insight? Why turn to science to bolster belief or faith in an essentially spiritual insight? Laszlo is clear that one of the hallmarks of science—and the main reason for trusting scientific knowledge—is that it is both empirical and experimental. Science tests its theories and thereby produces reliable, predictable, and practical knowledge.

I share Laszlo's inclination to trust and value science, and I too feel encouraged to know that the scientific evidence for the quantum vacuum conforms so well with the ancient idea of a universal or cosmic Akashic Record.

But why? Why should we trust that just because science says so, we are more likely to have a truer explanation of reality than we get from spiritual or mystical experience or insight? In certain "moods" or more intuitive states of mind, I find myself deeply trusting and valuing spiritual wisdom, accepting that it offers a profound understanding or insight into the ultimate nature of reality. Nevertheless, there is something deeply reassuring in knowing that science also reveals a similar understanding of the world.

One reason, I think, is that science is widely accepted in modern society as the legitimator, the ultimate arbiter, of what is really real. Yes, we know that science has its limitations and blind spots, but overall it does a good job of exploring our world and presenting us with *usable* and *repeatable* knowledge.

But how? What is it about science that enables it to produce such pragmatic and practical knowledge—knowledge that empowers us to change our world (for good or ill)? Well the most distinctive mark of science is not merely that it tests its theories, but that it tests by *measurement*. Science works because it uses a methodology that extracts information from the world by measuring it. And measurement removes guesswork. If done with precision and accuracy, it yields repeatable, reliable, reusable knowledge.

What does it mean to measure something? Basically, it is a process of assigning numbers to physical quantities by using a standard for comparison (for example, a ruler, or a scale). Science is a method for quantifying and measuring physical reality; equipped with such data, we are empowered to manipulate the world, to adapt it to our needs and desires. In short: we trust and value science because it works.

I think it comes down to this: We trust science because it possesses the tools to explore, measure, and explain happenings in the physical world—the world of things we need for surviving and thriving. From the perspective of day-to-day living, the physical world is the *real* world. It contains the objects that we can see, hear, smell, taste, and touch—the world revealed to us by our senses. We have very good reasons for putting a lot of epistemological weight on what we learn and know through our senses. Knowledge not based on sensory input tends to be dismissed in our culture as "speculative" or "imaginary." And if it is not tested, then why should we believe it tells us anything about the real world—why should we take it as "knowledge" at all? Science makes sense because it is based on what the senses reveal, and is tested by rigorous experiments.

Many scholars have previously attempted to bridge the two great traditions of science and spirituality. For example, in the 1970s Fritjof Capra's influential book *The Tao of Physics* pointed out remarkable parallels between quantum physics and the mystical insights of Taoism, Hinduism, and Buddhism. In that case, the unifying idea was universal *interconnectedness.* Now, Laszlo offers the beginnings of an explanation for that interconnectedness in the quantum vacuum or A-field. Here, the unifying idea is *information.* Like ripples on a pond, everything that happens in our world arises from (and *is*) interacting waves or patterns of energy and information in the A-field. These waves literally *in-form,* or give shape to, the manifest physical world. And, in Laszlo's theory, these waveforms are encoded holographically throughout the entire A-field. The A-field, thus, is a permanent and complete record or repository of *all* energy and information—of everything that has ever happened. It is also the *source* of everything that has ever happened, is happening, and will happen.

In *Science and the Akashic Field* Laszlo describes our world, the manifest reality we inhabit, as "the informed universe." It springs forth from the energy and information of the quantum vacuum or A-field, and is literally "in-formed" or seeded by templates stored there from prior universes that have completed their journey from big bang to catastrophic crunch (or some other inevitable eventual demise), and have returned all their energy and information to the quantum "void." We live in a "recycled" or "born-again" universe, a world composed of energy and information handed on to us by the A-field. Our universe is just one (very long) event in an ongoing series of born–evolving–dying–born-again universes. Collectively, all worlds spring from the mother of all universes that Laszlo calls the Metaverse. We could think of each universe—ours included—as an event that begins, evolves, and eventually ends within the overarching, eternal process we call "Cosmos."

Consciousness All the Way Down. Cosmos, Metaverse, A-field, quantum vacuum (all versions of the same ultimate feature of reality), and our actual universe are composed of energy and information at all levels of existence—and the accumulation of information is the basis for Laszlo's ToE. But he is also keenly aware that this cannot be a full theory of everything because neither energy nor information can explain or account for consciousness. Recognizing that consciousness cannot emerge from mindless energy or information, Laszlo acknowledges the panpsychist view (championed by philosopher Alfred North Whitehead in *Process and Reality*) that for his ToE to hold together coherently, consciousness must be intrinsic to the A-field, not something that later emerges from it. Consciousness, therefore, is inherent in the cosmos all along. Consciousness is part of the very fabric of being.

Thus, deep reality has a tri-aspect nature, consisting of Information, Consciousness, and Energy: I.C.E. For Laszlo, information is the missing link from previous attempts to come up with a true ToE. In this and other books, he presents a highly understandable and accessible account of the crucial role of information stored in the A-field as the explanatory foundation for a theory of everything. This is a great unifying, integrative insight.

Drawing on this insight, Laszlo offers a provocative overview and a masterful synthesis of knowledge at the frontiers of cosmology, physics, neurobiology, and consciousness studies. However, for his vision to catch fire in the imaginations and belief systems of millions (where paradigm shifts occur), we need a deeper understanding and appreciation of two intriguing ideas that are fundamental to Laszlo's work.

Interconnectedness and Hierarchy. To say that everything is connected to everything else is one of those phrases that is simultaneously both bland and sublime. Its deep truth is so easily obscured by its universality. At one level, it seems almost obvious: How could there be nothing separating any two things? If they were truly separated by a void, how would they ever interact or communicate from their separate bubble universes? (The real conundrum in physics and philosophy is not so much "action-at-a-distance" as "action-through-a-void.") If there was literally "nothing" separating them, then, of course, logically they would be connected. And if there is something in between, that's what connects them. So, to be part of the same uni-verse, everything, it seems, must be connected in some way. There really isn't any other option: everything is necessarily interconnected.

This is a key conclusion from Laszlo's model: at its deepest level, the physical universe is a *nonlocal* quantum field. And one of the "amazing" discoveries from experiments involving nonlocality is that, indeed, since the entire universe began as a minute quantum system, and since any "parts" that once belonged to a quantum system remain connected or *correlated* no matter how far apart they may be separated in space, then the entire universe must be connected at a very deep level.

But is such an insight really helpful? If it spurs us toward more ecologically minded perceptions and actions, then, yes, it may be valuable. However, for any such ecologically motivated actions to be practical, we need to know more: We need to know not just that everything is interconnected, but *how* everything is connected. We need to know the *nature* and *structure* of interconnection.

And this brings us to the concept of "hierarchy."

To many people, the term "hierarchy" is emotionally loaded. It smacks of authoritarian power structures. In some postmodern critiques hierarchy is associated with patriarchal power plays, the kind of intellectual and social paradigms that have led to the creation of military and industrial complexes that have so devastated human societies and natural environments. Hierarchy has had bad press, and much of it is deserved. But that is because the word has been co-opted and abused by some of the power structures that have claimed to be models of hierarchy.

Originally, the word "hierarchy" meant something quite different from the connotations it carries today. In the early middle ages, the word was used by the Church to characterize the relationships within ecclesiastical orders and between the clergy, the faithful, nature, and their God. Its original meaning referred to a "sacred relationship" among the celestial orders of angels, with seraphim and cherubim closest to God, followed by archangels and then angels. "Hierarchy" (from *hiero-*, meaning "sacred or holy," and *-arche*, meaning "governance or rule"), thus referred to the "sacred governance" between creatures and their divine creator.

Later, as the Church rigidified into dogmatism, with a power structure designed to enforce the "word of God" as interpreted by the

Church authorities, "hierarchy" lost its sense of sacred connection, and came to mean stratified levels of authority and the exercise of power. As a mechanism for asserting obedience, subjugation, and domination, this corrupted version of hierarchy became a natural model for armies to adopt. Today, the concept of hierarchy is most readily associated with chains of military command and with exploitative corporate employee relations. Hence its bad press.

However—no matter how justified we may be to bristle at the notion of hierarchy (as corrupted)—we should be careful not to lose sight of the usefulness of the uncorrupted concept for describing actual relations between natural systems. Whether or not we object to the implicit notion of authoritarian power structures, it is difficult to consistently deny that nature appears to be structured as levels of organization or complexity. Elementary particles give rise to atoms, atomic structures form molecules, which in turn form macromolecules such as proteins and DNA, which are the basis for living organelles and cells, which congregate and cooperate to form the profusion of living organisms populating the planet. To deny this structuring of life within life, built up from simpler inorganic constituents, is to deny the complex ecological relationships that have come about through evolution. Evolution, as a progressive complexification of matter and psychobiotic systems, is ostensibly a dynamic process of ever-increasing levels of complexity and organization. In this sense, in the sense of nested systems within systems, hierarchy is an accurate and appropriate description of nature.

What hierarchy does *not* mean, in this sense, is a one-way chain of command, with power residing either at the top (authoritarianism), or at the bottom (reductionism). If we picture nature's nested systems as circles within circles within circles, where the boundaries of all the circles are permeable, then hierarchy permits the flow of information and energy both up and down, and laterally, between systems at all levels. Hierarchy involves communication of information and energy through "upward causation," from lower-level (meaning less complex) systems to higher-level (meaning more complex and organized) systems, and "downward causation," from higher-level systems to their component parts; as well as horizontal causation (laterally between systems on the same level). In this systems view of hierarchy, power resides in the cooperative relationships between the various systems and their parts.

We really need to distinguish between two very different kinds of hierarchy (as writer and scholar Riane Eisler has shown). On the one hand we have *actualization hierarchies* that exist to optimize an organism's (or system's) potentials at all levels. In this kind of hierarchy, the whole serves the parts simultaneously as the parts contribute to the whole. Individual initiative is encouraged and enabled as long as it serves the healthy functioning of the greater whole (including the health of the parts). In return, healthy functioning of the whole generates more options for creative initiative from the parts. When we are healthy, this kind of actualization or *functional* hierarchy is happening every moment within our own bodies, sustaining and nurturing us. Not only do we want this kind of hierarchy—we *need* it. Without hierarchy, we'd all be undifferentiated mush!

By contrast, there are also *dysfunctional hierarchies,* which are based on force (or threats of force), and exist for the benefit of a privileged few at the top. These "hierarchies" rely on one-way communication of rules and orders to instill obedience. In such *dominator* hierarchies, higher levels use (or abuse) lower levels for selfish ends. They inhibit individual creativity and choice, and are ultimately and intrinsically pathological. In fact, these distorted versions are not true hierarchies at all; rather, they are unbalanced systems of coercion and domination.

Science writer and scholar Arthur Koestler coined a new term, "holon," to refer to the healthy systemic nature of the relationship between wholes and parts. Every whole is also a part of a greater system; and every part is a whole in relationship to its components. Holons are thus "Janus-faced" entities, facing both upward and downward (Janus was an ancient dual-faced Roman god who could see in opposite directions). Sensitive to the negative connotations of the historically loaded version of "hierarchy," Koestler proposed an alternative term, "holarchy," to capture the sense of multidirectional open-flow between nested systems.

Whether we prefer to adopt "hierarchy" or "holarchy," or some other term, the universe presents itself to us as a system composed of parts-within-wholes, of systems within systems, organized through time and evolution as interdependent levels of complexity. Each part, including you and me, is integral to the whole; and, in some holographic sense, each part is a microcosm of the greater macrocosm. Each part contains within itself the seed or template of the whole.

This has been the way of the world since the dawn of time. And that's the point: Hierarchy is the universal structure that embraces parts and wholes, and through the dynamics of evolution creates systems out of chaos. The entire universe, including us, is interconnected through *hierarchy.*

There is no system without chaos, just as there is no chaos without systems. Chaos and systems are embedded in each other all the way up and all the way down. We could say that the essential nature of the vacuum is "chaordic"—a complementarity of chaos and systemic order. They are like life and death, existence and the void. Just as the quantum "void" is a plenum, brim-full of potential for a universe bursting with evolution and creativity, so too chaos is both the source and sink of systems—systems forever exploring pathways of innovation, building up new forms of organization and information, and dissipating disorder (and information) back into the womb of quantum vacuum "chaos."

And it is just as likely that this process is never-ending as that it never had a beginning. Laszlo's idea of an eternal recycling of universes out of and back into the mother "Metaverse" explicitly proposes this—a truly never-ending story.

Consciousness in the Reenchanted Reality

Edgar Mitchell

One of the major discoveries of late twentieth-century science was evidence for the extension of quantum mechanics into macroscale physics from its origins in subatomic physics. As Ervin Laszlo points out, of particular importance is the extension of the major quantum properties of entanglement, coherence, nonlocality, and resonance into biology, consciousness studies, and beyond. Long sought understanding of exceedingly thorny problems in science may begin to yield to new analyses in coming decades using these concepts.

The four centuries following Rene Descartes' seventeenth-century conclusion of a mind/body duality saw the rise of modern science in the Western world, free from ecclesiastic dominance by the Inquisition. That is the up side. The understood caveat, of course, is that mind, consciousness, and spirit were shifted beyond the realm of the physical, and became appropriate for study within theology and philosophy, but not appropriate for natural science. Thus for most of the last four hundred years, scientists have framed theory and experiment to explain physical phenomena only. The result has been a materialistic science, devoid of inquiry into questions of deepest import to human interest. That is the down side.

The discovery, initiated by Max Planck, that light came in discrete packets that he called *quanta,* contemporary with Einstein's 1905 paper establishing that light possessed both wave and particle characteristics, set the stage for the subsequent emergence of quantum mechanics over the following two decades. But it was soon discovered that not only light, but all atomic scale matter exhibited the wave/particle duality, and that the act of measuring ("knowing") the physical characteristics of atomic and subatomic particles, such as position, momentum, and velocity, influenced the subsequent state of the particles. Consequently, the Cartesian conclusion that mind and matter did not directly interact was challenged by the experimental evidence of quantum mechanics.

Although this evidence had considerable philosophic and theological consequence, as a practical matter particle physicists continued to practice quantum physics for the next seventy years as an abstract and experimental, technical discipline devoid of major philosophic implication for the macroscale workaday world. Indeed, standard dogma has been that quantum mechanics only pertains to subatomic reality. Nearly a century later, high school students who may have been taught principles of Newtonian classical physics still have little or no understanding of quantum principles and their applications. Yet the formulas of quantum physics remain the most well tested and successful theories in modern science, having brought us nuclear understanding, and a host of technologies in communications, computation, and electronics. The attributes of quantum mechanics that set it apart from Newtonian mechanics are entanglement, quantum correlation, coherence, and nonlocality.

In *Science and the Akashic Field* and this book Laszlo reviews certain mental phenomena such as transpersonal mind-to-mind resonance and remote reception. It has been known since the early 1970s that these are *nonlocal* phenomena and cannot be shielded with electromagnetic means. Nonlocality is the strange aspect of quantum theory where particles in a process become entangled, and maintain a coherent correlation and an instantaneous communication even though one may be transported across the universe from the other. This suggests that the universe is "reenchanted": informationally interconnected in the mysterious way Einstein called "spooky action at a distance."

In 1994 Walter Schempp at the University of Seigen in Germany, while working to improve functional magnetic resonance imaging (fMRI) technology, discovered that if quantum emissions from matter were studied as a group phenomenon, rather than as individual photon emissions, they carried coherent information about the history of the object—*nonlocally and irreversibly.* (Most quantum processes are considered reversible.) He called his discovery Quantum Holography (QH) because it exhibits the coherent, distributed, entangled properties of laser holography, and exhibits nonlocality as well. An analogy to assist our understanding of the group phenomenon is the set of human fingerprints. If one studies a single small ridge on a person's finger, it does not tell us much about the individual, but the prints from all ten fingers, taken as whole, uniquely identify the individual. Similarly, studying the quantum emissions from matter as a group pattern provides unique information about that larger organization of matter. This discovery underscores Laszlo's claim that the role of information in the universe is of equal importance to that of energy itself. Our physical existence depends upon the condensed energy we call matter, but the ability of creatures in the universe to know anything about anything is based upon information.

For more than a century all types of psychic events have been denied validity by scientists as not having an explanation in science. But suddenly, with quantum holography, an information mechanism was discovered that seems to explain not only this group of well-known phenomena, but opens a new door to a line of investigation that offers the possibility of revolutionizing quantum theory by demonstrating that information is a fundamental player in the way Nature is organized at all scale sizes, and as an evolutionary mechanism. For example, it is colloquial in English to describe intuition as the "sixth sense," implying that it is a real sense, but its operation is more peripheral than that of the other five senses. But it now seems clear that—through the exchange of holographic information with the surrounding field Laszlo calls the A-field—an organism "learns" to adapt to the environment. In planetary evolution, the quantum holographic information is earlier and more fundamental than the five normal senses of evolved biological species. Thus intuition and instinct should more rightly be labeled the "first sense," as it is a more primitive attribute of our evolutionary nature than our other senses.

A number of phenomena reported and accepted as cultural lore, and some phenomena from mystical traditions, seem amenable to explanation with the A-field quantum holographic mechanism. The most promising candidates are: 1) passive perceptual phenomena, regardless of nomenclature, such as intuition, remote viewing, ESP, clairvoyance, clairaudience, (except for precognition, which requires a different theoretical approach); 2) significant revisions to basic theories of normal perception; 3) past-life regressions and reincarnation experiences; 4) the Akashic Record, which Laszlo considers a field phenomena and ascribes to the A-field as well; 5) the "twin phenomenon," the label attached to the experience of twins and other bonded individuals who frequently share perceptions and make similar choices even though separated by great distances; 6) acupuncture, subtle energies, and related therapies.

A common subjective experience is shared by people entering ancient structures, such as venerated cathedrals that have been in use for centuries. Those entering have a feeling of awe, hush, reverence, and peace that is palpable. What mechanism could cause that? The quantum holographic formalism associated with the A-field suggests a continuous exchange of information between people, objects, and their environment. Thus both the people and the cathedral are emitting an aura in the form of holographic information that becomes absorbed by the other. Presumably the years of reverence, hush, and awe in the psyche of worshippers is absorbed over time into the very structure of the building and is re-emitted into the space. The space is "conditioned" by the hordes who have visited it. It is interesting that one can visit the more mundane spaces of such a structure, say an office room, and the feeling present in the larger sanctuary is either vastly diminished or totally absent.

William Tiller of Stanford and his team have successfully demonstrated the "conditioning" of a space in a laboratory, and that certain properties of a material object can be altered by a coherent focused thought, as in meditation. In particular he has demonstrated that an "intention" such as "raise (or lower) the pH" of a sample of water can be effected by this coherent meditative procedure (pH is a measure of the acidity or alkalinity of a substance). He has further demonstrated that the intent can be stored in a conditioned electrical circuit and transported to a remote laboratory to effect the same result in that laboratory. This conditioning suggests a resonance between the perpetrator of the intention and the physical object, similar to the above-mentioned example of the cathedral.

To understand how this is taking place, Laszlo asks us to consider the example of the sea, where the waves created by objects interpenetrate and convey information on the objects that created them. We can also think of bats and dolphins. In order to locate their targets, they emit signals and record their return, similarly to a sonar or a reflected radar beam. Or think of reverberating organ pipes, and how pipes of different lengths create sounds of different frequencies. This is known as a resonant condition. In each case, to mathematically model the perception of the signal and its location with respect to the observer requires a concept called phase conjugation, which means creating a wave that is equal and opposite in phase to the original. In other words two equal waves traveling in opposite directions are said to be phase conjugate. Add them together and they create a standing wave, which results in a resonance. In this way we can model the principle that allows bats, dolphins, and whales to locate objects in their vicinity. It allows radar on ships and aircraft to locate objects in their field of view and place the image on a screen for the operator to see, or MRI machines to provide clear images inside the body.

An object emitting a quantum hologram and the receiver of that information must be in the relationship of phase conjugation in order for the information to be perceived. This is termed phase conjugate adaptive resonance *(pcar)* in regard to the quantum hologram. That is to say, the receiver, say a brain, must in effect create a virtual signal that is phase conjugate to the incoming wave. It is thought that the microtubules in the brain/body function in exactly this way in order to perceive and locate objects in the external world. The incoming light-wave from an object in the field of view, and the quantum holographic emissions from that object, are placed in *pcar* in order for the object to be perceived by the optical nerve. The same is true regarding a sound being heard by the auditory system as emitting from the point of origin. Try the experiment of snapping your fingers and note that the sound is seemingly at the location of the fingers, not at a location in the brain. To model this physically and mathematically requires the concept of phase conjugation and resonance, not just a single wave traveling toward the body/brain, as we have tended to believe.

We have been taught that we perceive the world in three dimensions because we have two eyes set apart. This permits binocular vision and perception of three dimensions (depth perception). Close one eye, the 3D images are still there. Thus the binocular analogy is incomplete. But a phase conjugate model does describe the phenomenon correctly. On the other hand, if you close your eyes and imagine a familiar scene, the image appears to be in the frontal lobe of the brain and does not have the vivid 3D characteristic of vision.

The mathematical formalism used to model the quantum holographic phenomenon dates back to the work of Werner Heisenberg and Paul Dirac in the 1920s. The Heisenberg "nilpotent Lie group" algebra and the Dirac formulations of quantum mechanics are standard mathematical fare in quantum mechanics. The nilpotent formulation focuses on the information content of a phenomenon rather than on the amount of energy utilized. To illustrate: an underwater earthquake can produce a tsunami that moves huge amounts of water across the ocean very rapidly and has energy for great destruction. The sound of two rocks clicked together underwater, or a dolphin's signal, moves even more rapidly through the water, but not necessarily through great distances. It carries information but imparts an insignificant amount of energy. This leads to the concept of "nilpotence," a mathematical formulation that recognizes the distinction between an energy-laden process and an informational process.

It is important to recognize that in our attempts to understand natural phenomena through mathematical models, it is seldom possible for a single theory and the equations that express the theory to account for all effects attendant to an action, such as the energy/information example. The amount of energy required to convey a huge amount of information is often insignificant, and might be lost in an equation concerned with energy transfer. And this leads us to a discussion of zero-point energy (ZPE).

Zero-Point Energy. Laszlo cites my claim that in higher states of awareness every cell of the body coherently resonates with "the holographically embedded information in the quantum zero-point energy field." Let me discuss here the considerations that have led me to this claim.

It was the belief for centuries that the "aether" was the substance occupying free space. Likewise the debate as to whether light was particle or wave existed for centuries, given that a wave requires a medium to propagate. The Michelson/Morley experiment, at the turn of the twentieth century, determined that the speed of light in space is constant regardless of the direction of travel with respect to the Earth's motion. However, a velocity differential was expected, depending on whether the speed of light measurement was taken parallel or perpendicular to the motion of the Earth (as a result of the resistance of the ether). This expected difference was found not to exist. The experiment led (erroneously) to the interpretation that space was completely empty, and it did not resolve the question of what medium supported the wave aspect of light.

Of course, it was quickly realized that the frequency (and therefore the color) of light, and its propagation speed, change with the relative motion between the source and the perception of light waves; and that the frequency characteristics of light emitted by a star are determined by the particular nuclear interactions taking place in that star. Modern studies show that light does travel at different speeds in different mediums, such as fiber optic cables and planetary atmospheres. This has led to a conundrum about the properties of free space that determine the speed of light. Does free space have measurably different characteristics in the vicinity of large planets? The answer seems, "yes."

The study of black body radiation in the same period brought the understanding that even at a temperature of zero degrees Kelvin (which is absolute zero), and in a vacuum chamber, matter emits photons. So the evidence that space is not truly empty even though it has no discernable molecules was observed very early in the investigation of quantum phenomena. However, Einstein's theory of general relativity assumed space to be empty and accounted for the bending of light around large stellar objects by the warping of the space/time continuum. In the emerging view—the view of a reenchanted cosmos—space is not empty and the term "zero-point field" (or "nuether") is often used to describe the interstellar medium that permeates it. It is believed by many physicists that the zero-point field contains huge amounts of untapped energy, representing a source of unlimited potential for further investigation and future utilization. But to date, attempts to tap into this potential have yielded very limited success. This is one area for frontier science.

Many books have been written on the various attempts to explain and understand instantaneous action at a distance. However, the precise mechanism of this phenomenon remains elusive. The Alain Aspect experiments of 1982 in France clearly demonstrated that the nonlocal phenomenon is real and valid. It is clear that, as this book affirms, the zero-point field is a medium that is involved with the phenomenon of nonlocal propagation of quantum information and quantum resonance. We will leave the issue of the precise mechanism of nonlocality as one for future resolution. However, the A-field—the nonlocal quantum holographic phenomenon residing in the zero-point field—already helps to clarify a number of mysteries.

As Laszlo points out, the quantum hologram information offers an explanation for evidence of claims of reincarnation, and a mechanism for past-life recall under hypnosis. This is not to say that conventional religious beliefs and interpretations about reincarnation are validated by this mechanism. They are not. Rather this mechanism provides a physical foundation for the experience of a previous life preserved through its quantum holographic information in the nonlocal zero-point A-field. It is recovered by a subsequent living being in the manner that any psychic information is recovered. The quantum hologram, being an information mechanism, does not speak to the question of the survival of consciousness. That phenomenon, espoused by most religions, should it be a valid understanding, will require a level of proof not yet in evidence.

The Nature of Consciousness. The truly tough question about consciousness is to understand its exact nature. Idealist philosophy, espoused by most major religions, holds that consciousness is fundamental and that the physical universe is a creation of an omniscient primordial mind. Science, on the other hand, claims that mind and consciousness are products of the organization of matter; they are an epiphenomenon of material existence. Definitive experimental evidence to validate either of these interpretations is not yet available. But, the same as Laszlo, this writer leans toward a cosmology that allows consciousness and matter to evolve together toward ever greater levels of complexity, beginning with the observed phenomenon of the quantum correlation of particles as primitive awareness at the most basic level of Nature. The quantum holographic discovery lends credence to this view. As the complexity of material form evolves, emitting and reabsorbing quantum information, so also does the complexity of awareness, perception, and consciousness.

The knowledge and technologies that are most certain to arise from these frontier concepts will certainly be as astounding to us as electricity, automobiles, and airplanes were a scant hundred years ago. Physics, biology, cosmology, and medical knowledge, to name but a few, will change significantly in this century.

The Akashic Field and the Dilemmas of Modern Consciousness Research

Stanislav Grof

In the course of the twentieth century, various disciplines of modern science amassed an extraordinary array of observations, which could not be accounted for and adequately explained in terms of the monistic materialistic worldview and within the context of what Fritjof Capra called the Newtonian-Cartesian paradigm. These paradigm-breaking "anomalous phenomena" came from a wide range of fields, from astrophysics, quantum-relativistic physics, and chemistry, to biology, anthropology, thanatology, parapsychology, and psychology.

Pioneering scientists from various scientific disciplines made more or less successful attempts to tackle the formidable conceptual problems presented by the anomalous data. They formulated theories, which suggested revolutionary new ways of looking at the recalcitrant problems they were facing in their respective fields. Over time, a radically different understanding of reality and of human nature started to come into view, which is usually referred to as the "new paradigm" in science. However, this new perspective was a mosaic consisting of impressive but disjointed pieces. It lacked "conceptual glue," a unifying vision that would seamlessly integrate the individual contributions into a comprehensive overarching "theory of everything."

Ken Wilber expressed in his writings the need for an integral theory of everything and outlined what such a theory should look like. However, the credit for actually creating such a theory goes to Ervin Laszlo, arguably the world's greatest system theorist and interdisciplinary scientist and philosopher. Laszlo, who is of Hungarian origin and currently lives in Italy, is a multifaceted individual with a range of interests and talents reminiscent of the great figures of the Renaissance. He achieved international fame as a child prodigy and in his teens performed as a concert pianist. A few years later, he turned to science and philosophy, beginning his lifetime search for the understanding of human nature and of the nature of reality. In addition, he has played an important role in addressing urgent sociopolitical problems of our times.

In an intellectual *tour de force* and a series of books, Laszlo explored a wide range of disciplines, including astrophysics, quantum-relativistic physics, biology, and psychology. He pointed out a wide range of phenomena, paradoxical observations, and paradigmatic challenges, for which these disciplines had no explanations. Drawing on advances of hard sciences and on mathematics, he offered a solution to the current paradoxes in Western science, which transcends the boundaries of individual disciplines. Laszlo achieved this by formulating his "connectivity hypothesis," the main cornerstone of which is the existence of what he called the "psi-field" and, more recently, the "Akashic field." He describes it as a subquantum field, which holds a holographic record of all the events that have happened in the phenomenal world.

In this comment, I shall focus on the groundbreaking contributions Ervin Laszlo's work has made to two areas that are closest to my heart—consciousness research and transpersonal psychology; the latter is a field of psychology of which I am a co-founder.

In the last fifty years, my primary area of interest and my lifetime passion has been the study of an important subcategory of non-ordinary states of consciousness, which I call holotropic (that is, oriented toward wholeness) states. This group includes the states which shamans experience in their initiatory crises and use in their healing practice, those that native people experience in rites of passage and healing ceremonies, as well as those that were described in the reports of neophytes who underwent initiation in the ancient mysteries of death and rebirth. Holotropic states also occur in the systematic spiritual practices of yogis, Buddhists, Taoists, and Sufis, and they have been described by mystics of all countries and historical periods. Modern psychiatrists and therapists encounter these states in psychedelic therapy, in deep experiential work without psychoactive substances (primal therapy, rebirthing, holotropic breathwork), and in the work with people in psychospiritual crises ("spiritual emergencies").

My interest in non-ordinary states of consciousness began almost fifty years ago, when a powerful experience lasting only several hours of clock-time profoundly changed my personal and professional life. As a young psychiatric resident, only a few months after my graduation from medical school, I volunteered for an experiment with LSD, a substance with remarkable psychoactive properties, which had been serendipitously discovered by the Swiss chemist Albert Hofmann in the Sandoz pharmaceutical laboratories in Basel, Switzerland. This session, particularly its culmination period, during which I had an overwhelming and indescribable experience of cosmic consciousness, awakened in me an intense lifelong interest in holotropic states.

Since that time, most of my clinical and research activities have consisted of systematic exploration of the therapeutic, transformative, and evolutionary potential of holotropic states. The five decades that I have dedicated to consciousness research have been for me an extraordinary adventure of discovery and self-discovery. I spent approximately half of this time conducting therapy with psychedelic substances, first in Czechoslovakia, in the Psychiatric Research Institute in Prague, and then in the United States, at the Maryland Psychiatric Research Center in Baltimore, where I participated in the last surviving official American psychedelic research program. Since 1975, I have worked with holotropic breathwork, a powerful method of therapy and self-exploration that I have developed jointly with my wife Christina. Over the years, we have also supported many people undergoing spontaneous psychospiritual crises, or "spiritual emergencies" as Christina and I call them.

The common denominator of these three situations is that they involve holotropic states of consciousness. In psychedelic therapy, these states are induced by administration of psychoactive substances, such as LSD, psilocybin, mescaline, and tryptamine or amphetamine derivatives. In holotropic breathwork, consciousness is changed by a combination of faster breathing, evocative music, and energy-releasing bodywork. In spiritual emergencies, holotropic states occur spontaneously, in the middle of everyday life, and their cause is usually unknown.

In addition, I have been more peripherally involved in many disciplines, which are, more or less directly, related to holotropic states of consciousness. I have participated in sacred ceremonies of native cultures in different parts of the world, involving use of psychedelic plants, such as peyote, sacred mushrooms, ayahuasca, and kava kava. My wife and I have had many encounters with North American, Mexican, and South American shamans, and exchanged information with many anthropologists. I have also had extensive contact with representatives of various spiritual disciplines, including Vipassana, Zen, and Vajrayana Buddhism, Siddha Yoga, Tantra, and the Christian Benedictine order.

Another area that has received much of my attention has been thanatology, the young discipline studying near-death experiences and the psychological and spiritual aspects of death and dying. I participated in the late 1960s and early 1970s in a large research project studying the effects of psychedelic therapy in individuals dying of cancer. I should also add that I have had the privilege of personal acquaintance and experience with some of the great psychics and parapsychologists of our era, pioneers of laboratory consciousness research, and therapists who developed and practiced powerful forms of experiential therapy, which induce holotropic states of consciousness.

My initial encounter with holotropic states was very difficult and intellectually, as well as emotionally, challenging. In the early years of my laboratory and clinical research with psychedelics, I was daily bombarded with experiences and observations, for which my medical and psychiatric training had not prepared me. As a matter of fact, I was experiencing and seeing things which, in the context of the scientific worldview I was brought up with, were considered impossible and were not supposed to happen. And yet, those obviously impossible things were happening all the time.

Instead of talking in abstract terms, let me briefly review the most important paradigm challenges that emerged from this research. I came into clinical work with psychedelics as a beginning psychiatrist, prepared by my medical education and equipped with passing knowledge of Freudian psychoanalysis. In serial psychedelic sessions, every single patient we worked with sooner or later transcended the narrow conceptual framework of mainstream psychiatry, which is limited to postnatal biography and to the Freudian individual unconscious. In my work with LSD and psilocybin, I witnessed practically daily deep and authentic age regression in my clients, which took them far beyond the map I had inherited from my teachers, to memories of biological birth and of prenatal existence. The same happened very early in my own psychedelic experiences. For about three years, I carefully charted the experiential territories traversed in psychedelic sessions, creating a vastly expanded cartography of the psyche to accommodate the new data.

The new map that emerged from this work shared with academic psychology and psychiatry the biographical-recollective domain and the Freudian individual unconscious. However, while in the mainstream model of the psyche these two regions represent the totality of the psyche, my new cartography contained two large and important additional domains. Both of these domains were transbiographical in nature, in the sense that they lay beyond (or beneath) the realm of conscious and unconscious contents related to postnatal biography. I called the first of these two new domains *perinatal,* which reflects its close connection with biological birth, and the second *transpersonal,* or reaching beyond the personal identity as it is usually defined.

This expanded cartography of the psyche was not really entirely new. Although it emerged independently from my own research, it represented a synthesis of the perspectives of various schools of depth psychology known from the history of psychoanalysis. It included, besides some of Freud's original concepts, important revisions proposed by famous psychoanalytic renegades. Thus, Otto Rank described in his pioneering book *The Trauma of Birth* the existence of the perinatal unconscious and emphasized its importance for psychology and psychotherapy. Psychoanalysts Lietaert Peerbolte and Nandor Fodor independently confirmed Rank's ideas about the importance of the trauma of birth and prenatal dynamics for psychology and psychotherapy.

In his 1949 book Wilhelm Reich discovered the powerful energies stored in the psyche and responsible for what he called the character armor. In 1961 and 1970 he discussed the important role they play in a wide range of phenomena from psychoneuroses and psychosomatic disorders to sociopolitical movements. Another of Freud's followers, Sandor Ferenczi, seriously considered, in his 1938 essay *Thalassa,* the possibility that the deep unconscious harbored memories of life in the primeval ocean. By far the most radical revision was introduced by C. G. Jung, with his discovery of the collective unconscious and its governing principles, the archetypes. As I described in 1985 in another context, the ideas of the above pioneers of depth psychology had to be significantly modified in the light of the observations from holotropic states before they could be integrated into the new cartography.

The rich and complex material originating on the *perinatal level of the unconscious* appeared in psychedelic sessions in several typical clusters, matrices, or experiential patterns. It soon became obvious that these categories of experiences were closely related to the consecutive stages of the biological birth process and to the experiences of the child in the perinatal period. I have coined for them the term *Basic Perinatal Matrices (BPMs).*

The second major domain that has to be added to mainstream psychiatry's cartography of the human psyche when we work with holotropic states is now known under the name *transpersonal,* meaning literally "beyond the personal" or "transcending the personal." The experiences that originate on this level involve transcendence of the usual boundaries of the body/ego and of the limitations of three-dimensional space and linear time that restrict our perception of the world in the ordinary state of consciousness. Transpersonal motifs can appear in holotropic states in various combinations with perinatal elements or independently of them. The transpersonal realm is the source of a wide range of anomalous phenomena, which present serious challenges, not only to current conceptual frameworks of psychology and psychiatry, but also to the monistic, materialistic philosophy of modern science.

Transpersonal experiences can best be defined by describing how they differ from our everyday experience of ourselves and of the world. In the ordinary or "normal" state of consciousness, we experience ourselves as material objects contained within the boundaries of our skin and operating in a world with Newtonian characteristics. The American writer and philosopher Alan Watts referred to this experience of oneself as identifying with the "skin-encapsulated ego." Our perception of the environment is restricted by the physiological limitations of our sensory organs and by physical characteristics of the environment.

We cannot see objects from which we are separated by a solid wall, ships that are beyond the horizon, or the surface of the other side of the moon. If we are in Prague, we cannot hear what our friends are talking about in San Francisco. We cannot feel the softness of the lambskin unless the surface of our body is in direct contact with it. In addition, we can experience vividly and with all our senses only the events that are happening in the present moment. We can recall the past and anticipate future events, fantasize about them, or try to use various methods to predict them. However, these are very different experiences from an immediate and direct sensory perception of what is happening in the present moment. In transpersonal states of consciousness, none of these limitations are absolute; any of them can be transcended.

Transpersonal experiences can be divided into three large categories. The first of these primarily involves transcendence of the usual spatial barriers, or of the limitations of the "skin-encapsulated ego." Here belong experiences of merging with another person into a state that can be called "dual unity," assuming the identity of another person, or identifying with the consciousness of an entire group of people (such as all mothers of the world, the entire population of India, or all the inmates of concentration camps). In the extremes, it is even possible to experience an extension of consciousness, which is so enormous that it seems to encompass all of humanity, the entire human species. Experiences of this kind have been repeatedly described in the spiritual literature of the world.

In a similar way, one can transcend the limits of the specifically human experience and identify with the consciousness of various animals, plants, or even a form of consciousness that seems to be associated with inorganic objects and processes. In rare instances, it is possible to experience consciousness of the entire biosphere, of our planet, or the entire material universe. Incredible and absurd as it might seem to a Westerner subscribing to the worldview created by materialistic science, these experiences suggest that everything that we can experience in our everyday state of consciousness as an object has a corresponding subjective representation in the holotropic states. It is as if everything in the universe has its objective and subjective aspect, just the way it is described in the great spiritual philosophies of the East. For example, in Hinduism all phenomenal worlds are seen as the divine play of Absolute Consciousness or Brahman *(lila),* in Taoism all elements of material reality are described as transformations of the Tao, and so on.

The second category of transpersonal experiences is characterized primarily by the overcoming of temporal rather than spatial boundaries, that is, by transcendence of linear time. I have found that there is the possibility of a vivid reliving of important memories from infancy, childhood, birth, and prenatal existence. In holotropic states, this historical regression can continue farther and involve what appears to be authentic experiential identification with the sperm and the ovum at the time of conception on the level of cellular consciousness. But the experiential travel back in time does not stop even there and can continue to episodes from the lives of one's human or animal ancestors, or even those that seem to be coming from the racial and collective unconscious as it was described by C. G. Jung. Very frequently, the experiences that seem to be happening in other cultures and historical periods are associated with a sense of personal remembering. People then talk about reliving of memories from past lives, from previous incarnations. These observations throw a new light on the problem of reincarnation, a concept of extreme spiritual and cultural importance, which has been pooh-poohed and ridiculed by materialistic science.

The idea of karma and reincarnation represents a cornerstone of Hinduism, Buddhism, Jainism, Sikhism, Zoroastrianism, Tibetan Vajrayana Buddhism, and Taoism. Similar ideas can be found in such geographically, historically, and culturally diverse groups as various African tribes, native Americans, pre-Columbian cultures, the Hawaiian kahunas, practitioners of the Brazilian umbanda, the Gauls, and the Druids. In ancient Greece, several important schools of thought subscribed to it. Among them were the Pythagoreans, the Orphics, and the Platonists. This doctrine was also adopted by the Essenes, the Pharisees, the Karaites, and other Jewish and semi-Jewish groups, and it formed an important part of the cabalistic theology of medieval Jewry. It was also held by the Neoplatonists and Gnostics.

The content of the transpersonal experiences described so far reflects various aspects of the material world and events happening at specific points of space and time. These experiences involved elements of the everyday familiar reality: other people, animals, plants, and materials. What was surprising about them was not their content, but the fact that we could witness or experientially identify with something that is not ordinarily accessible to our experience. We know that there are pregnant whales in the world, but we should not be able to have an authentic experience of being one. The fact that the French revolution took place is readily acceptable, but we should not be able to have a vivid experience of actually being there and dying on the barricades of Paris. We know that there are many things happening in the world in places where we are not present, but it is usually considered impossible to experience something that is happening in a remote location without the mediation of a TV camera, a satellite, and a TV set. We may also be surprised to find consciousness associated with lower animals, plants, and with inorganic nature.

However, the third category of transpersonal experiences is even stranger. Here consciousness seems to extend into realms and dimensions that the Western industrial culture does not consider to be "real." Here belong numerous visions of archetypal beings and mythological landscapes, encounters or even identification with deities and demons of various cultures, and communication with discarnate beings, spirit guides, suprahuman entities, extraterrestrials, and inhabitants of parallel universes. In its farther reaches, individual consciousness can identify with cosmic consciousness or the Universal Mind known under many different names: Brahman, Buddha, the Cosmic Christ, Keter, Allah, the Tao, the Great Spirit, and many others. The ultimate of all experiences appears to be identification with the Supracosmic and Metacosmic Void, the mysterious and primordial emptiness and nothingness that is conscious of itself and is the ultimate cradle of all existence. It has no concrete content, yet it contains all there is in a germinal and potential form.

Transpersonal experiences have many strange characteristics that shatter the most fundamental metaphysical assumptions of the Newtonian-Cartesian paradigm and of the monistic, materialistic worldview. Researchers who have studied and/or personally experienced these fascinating phenomena realize that the attempts of mainstream science to dismiss them as irrelevant products of human fantasy and imagination or as hallucinations—erratic products of pathological processes in the brain—are naive and inadequate. Any unbiased study of the transpersonal domain of the psyche has to come to the conclusion that the observations represent a critical challenge not only for psychiatry and psychology, but for the entire philosophy of Western science.

Although transpersonal experiences occur in the process of deep individual self-exploration, it is not possible to interpret them simply as intrapsychic phenomena in the conventional sense. On the one hand, they appear on the same experiential continuum as the biographical and perinatal experiences and are coming from within the individual psyche, in the sense that they are obtained by introspection. On the other hand, they seem to be tapping directly, without mediation of the senses, into sources of information that are clearly far beyond the conventional reach of the individual. Somewhere on the perinatal level of the psyche, a strange flip seems to occur and, what was up to that point deep intrapsychic probing, starts rendering experiences of the universe at large obtained by extrasensory means. Some people have compared this to an "experiential Moebius strip," since it becomes impossible to say what is inside and what is outside.

These observations indicate that we can obtain information about the universe in two radically different ways. Besides the conventional possibility of learning through sensory perception, and analysis and synthesis of the data, we can also find out about various aspects of the world by direct identification with them in a holotropic state of consciousness. Each of us thus appears to be a microcosm containing, in a holographic way, the information about the entire macrocosm. In the mystical traditions, this was expressed by such phrases as: "as above so below" or "as without, so within." In the past, this basic tenet of esoteric schools, such as Tantra, the Hermetic tradition, Gnosticism, and Kabbalah, appeared to be an absurd confusion of the relationship between the part and the whole and a violation of Aristotelian logic. In the second half of the twentieth century, this claim received unexpected scientific support in the discovery of the principles operating in optical holography.

The reports of subjects who have experienced episodes of embryonal existence, the moment of conception, and elements of cellular, tissue, and organ consciousness, abound in medically accurate insights into the anatomical, physiological, and biochemical aspects of the processes involved. Similarly, ancestral, racial, and collective memories and past incarnation experiences frequently provide very specific details about architecture, costumes, weapons, art forms, social structure, and religious and ritual practices of the culture and historical period involved, or even concrete historical events. People who experienced phylogenetic experiences or identification with existing life forms not only found them unusually authentic and convincing, but often acquired in the process extraordinary insights concerning animal psychology, ethology, specific habits, or unusual reproductive cycles. In some instances, this was accompanied by archaic muscular innervations not characteristic for humans, or even such complex behaviors as enactment of a courtship dance.

The scientific and philosophical challenges associated with the already described observations are further augmented by the fact that transpersonal experiences correctly reflecting the material world often appear on the same continuum and intimately interwoven with elements from the mythological world, which the Western industrial world does not consider to be ontologically real. Here belong experiences involving deities and demons from various cultures, abodes of the Beyond such as heavens and paradises, and legendary or fairy-tale sequences. For example, we can experience communication with Jesus, Virgin Mary, or the Devil, have a shattering encounter with the Hindu goddess Kali, or identify with the dancing Shiva.

In holotropic states, it is also possible to visit various mythological realms, such as Shiva's heaven, the paradise of the Aztec rain god Tlaloc, the Sumerian underworld, or one of the Buddhist hot hells. Even these episodes can impart accurate new information about religious symbolism and mythical motifs that were previously unknown to the person involved. Observations of this kind confirm C. G. Jung's idea that, besides the Freudian individual unconscious, we can also gain access to the collective unconscious that contains the cultural heritage of all humanity. Although these mythic elements are accessed intrapsychically, in a process of introspection, they have objective existence, are ontologically real.

To distinguish transpersonal experiences from imaginary products of individual fantasy, Jungians refer to this domain as *imaginal.* French scholar, philosopher, and mystic Henri Corbin, who first used the term *mundus imaginalis,* was inspired by his 2000 study of Islamic mystical literature. In Islamic theosophy, the imaginal world where everything existing in the sensory world has its analogue is called *'alam a mithal,* or "the eighth climate," to distinguish it from the seven climates, or regions, of traditional Islamic geography. It possesses extension and dimensions, forms and colors, but these are not perceptible to the senses, as they are when they are properties of physical bodies. However, this realm is in every respect as fully ontologically real as the material world perceived by the senses.

It is not an easy task to convey in a few sentences conclusions from daily observations made in the course of almost half a century of research on holotropic states of consciousness, and to make this statement believable. It is not realistic to expect that a few generalizations would be able to override the deeply culturally ingrained worldview of those readers who are not familiar with the transpersonal dimension and who cannot relate what I say to their personal experience. Although I myself had the opportunity to observe closely many hundreds of other people experiencing holotropic states and to listen to their accounts, it took me years to fully absorb the impact of the cognitive shock involved. The most convincing evidence for the validity of the astonishing and incredible new data did not come from observations of others, but from deep personal experience.

The recent revolutionary advances in science have been welcomed by transpersonal circles as significant conceptual support for transpersonal psychology. On the one hand, they undermined various aspects of the seemingly monolithic structure of the traditional materialistic worldview with which transpersonal psychology was in principle incompatible. On the other hand, they often brought new insights, which seemed to provide supportive evidence for various partial claims of the pioneers of the transpersonal perspective. However, what was still missing was a large integrative vision, the "conceptual glue" that would weave all these pieces into a comprehensive tapestry of ideas about consciousness, psyche, and human nature, one that could also be reconciled with the revolutionary findings of other scientific disciplines about the nature of reality.

It is Ervin Laszlo's work that has turned out to be the "Rosetta stone," which the pioneers of consciousness research and transpersonal psychology sought after. His contributions represent a quantum leap in the conceptual evolution and revolution described above. In a series of works, such as *The Creative Cosmos, The Interconnected Universe, The Whispering Pond, The Connectivity Hypothesis,* and most recently, *Science and the Akashic Field,* Laszlo reviewed the major theories that had attempted to solve the puzzles and paradoxes presented by the "anomalous phenomena"—the work of David Bohm, Karl Pribram, Rupert Sheldrake, Ilya Prigogine, and others. He showed the strength and the weaknesses of these theories and offered an elegant comprehensive metatheory, which addresses the unsolved problems in a number of disciplines.

The key element of this theory of everything, Laszlo's concept of the psi-field—also called the Akashic or A-field—a subquantum field containing the holographic record of everything that happens in the universe (or possibly the Kosmos as understood by ancient Greeks), certainly accounts for the otherwise baffling problems encountered by modern consciousness research and transpersonal psychology. As the title of his previous book—*Science and the Akashic Field*—suggests, Ervin Laszlo has not only been able to formulate a unifying conceptual framework for a number of scientific disciplines, but also to create a bridge that connects the best of hard science quite explicitly to transpersonal psychology and to the great spiritual traditions of the world.

Ervin Laszlo's creative interdisciplinary synthesis is by any standards an extraordinary intellectual achievement that deserves a prominent place in the history of science. I firmly believe that his work will in the near future be recognized and acknowledged as an important cornerstone of the scientific worldview of the future.

Akasha and Consciousness

Peter Russell

Ervin Laszlo has proposed that the virtual energy field known as the quantum vacuum, or zero-point field, corresponds to what Indian teachings have called Akasha, the source of everything that exists, and in which the memory of the cosmos is encoded. I would like to take his reasoning a step further and suggest that the nature of this ultimate source is consciousness itself, nothing more and nothing less.

Again we find this idea is not new. In the Upanishads, *Brahman,* the source of the cosmos (literally, "that from which everything grows"), is held to be equal to *Atman* ("that which shines"), the essence of consciousness. And in the opening lines of *The Dhammapada,* the Buddha declares that "All phenomena are preceded by mind, made by mind, and ruled by mind."

Such a view, though widespread in many metaphysical systems, is completely foreign to the current scientific worldview. The world we see is so obviously material in nature; any suggestion that it might have more in common with mind is quickly rejected as having "no basis in reality." However, when we consider this alternative worldview more closely, it turns out that it is not in conflict with any of the findings of modern science—only with its presuppositions. Furthermore, it leads to a picture of the cosmos that is even more enchanted.

ALL IN THE MIND

The key to this alternative view is the fact that all our experiences—all our perceptions, sensations, dreams, thoughts, and feelings—are forms appearing in consciousness. It doesn't always seem that way. When I see a tree it seems as if I am seeing the tree directly. But science tells us something completely different is happening. Light entering the eye triggers chemical reactions in the retina; these produce electro-chemical impulses, which travel along nerve fibers to the brain. The brain analyzes the data it receives, and then creates its own picture of what is "out there." I then have the experience of seeing a tree. But what I am actually experiencing is not the tree itself, only the image that appears in the mind. This is true of everything I experience. Everything we know, perceive, and imagine, every color, sound, sensation, every thought and every feeling, is a form appearing in the mind. It is all an "in-forming" of consciousness.

The idea that we never experience the physical world directly has intrigued many philosophers. Most notable was the eighteenth-century German philosopher Immanuel Kant, who drew a clear distinction between the form appearing in the mind—which he called the *phenomenon* (a Greek word meaning "that which appears to be")—and the world that gives rise to this perception, which he called the *noumenon* (meaning "that which is apprehended"). All we know, Kant insisted, is the phenomenon. The noumenon, the "thing-in-itself," remains forever beyond our knowing.

Unlike some of his predecessors, Kant was not suggesting that the perceptual reality is the *only* reality. Irish theologian Bishop Berkeley had likewise argued that we know only our perceptions. He then concluded that nothing exists apart from our perceptions, which forced him into the difficult position of having to explain what happened to the world when no one was perceiving it. Kant held that there *is* an underlying reality, but we never know it directly. All we can ever know of it is the form that appears in the mind—our mental model of what is "out there."

It is sometimes said that our model of reality is an illusion, but that is misleading. It may all be an appearance in the mind, but it is nonetheless real—the only reality we ever know. The illusion comes when we confuse the reality we experience with the physical reality, the thing-in-itself. The Vedantic philosophers of ancient India spoke of this confusion as *maya*. Often translated as "illusion" (a false perception of the world), maya is better interpreted as "delusion" (a false belief about the world). We suffer a delusion when we believe the images in our minds *are* the external world. We deceive ourselves when we think that the tree we see *is* the tree itself.

The tree itself is a physical object, constructed from physical matter: molecules, atoms, subatomic particles. But from what is the image in the mind constructed? Clearly it is not constructed from physical matter. A perceptual image is composed of the same "stuff" as our dreams, thoughts, and feelings, and we would not say that these are created from physical atoms or molecules. (There might or might not be a corresponding physical activity in the brain, but what I am concerned with here is the substance of the image itself.) So what is the mental substance from which all our experiences are formed?

The English language does not have a good word for this mental essence. In Sanskrit, the word *chitta,* often translated as "consciousness," carries the meaning of mental substance, and is sometimes translated as "mindstuff." It is that which takes on the mental forms of images, sounds, sensations, thoughts, and feelings. They are made of "mindstuff" rather than "matterstuff."

Mindstuff, or chitta, has the potential to take on the form of every possible experience: everything that I, or anyone else, could possibly experience in life; every experience of every being, on this planet, or any other sentient being, anywhere in the cosmos. In this respect consciousness has infinite potential. In the words of Maharishi Mahesh Yogi, "Consciousness is the field of all possibilities."

This aspect of consciousness can be likened to the light from a film projector. The projector shines light onto a screen, modifying the light so as to produce one of an infinity of possible images. These images are like the perceptions, sensations, dreams, memories, thoughts, and feelings that we experience—the forms arising in consciousness. The light itself, without which no images would be possible, corresponds to this ability of consciousness to take on form.

We know all the images on a movie screen are composed of light, but we are not usually aware of the light itself; our attention is caught up in the images that appear and the stories they tell. In much the same way, we know we are conscious, but we are usually aware only of the many different perceptions, thoughts, and feelings that appear in the mind. We are seldom aware of consciousness itself.

All phenomena are projections in the mind.

THE THIRD KARMAPA

NO MATTER?

Although we may not know the external world directly, we can draw conclusions from our experience as to what it might be like. This, in essence, has been the focus of our scientific endeavors. Scientists have sought to understand the functioning of the world around us, and draw conclusions about its true nature.

To the surprise of many, the world "out there" has turned out to be quite unlike our experience of it. Consider our experience of the color green. In the physical world there is light of a certain frequency, but the light itself is not green. Nor are the electrical impulses that are transmitted from the eye to the brain. No color exists there. The green we see is a quality appearing in the mind in response to this frequency of light. It exists only as a subjective experience in the mind.

The same is true of sound. I hear the music of a violin, but the sound I hear is a quality appearing in the mind. There is no sound as such in the external world, just vibrating air molecules. The smell of a rose does not exist without an experiencing mind, just molecules of a certain shape.

The same is also true of the solidness we experience in matter. Our experience of the world is certainly one of solidness, so we assume that the "thing in itself" must be equally solid. For two thousand years it was believed that atoms were tiny solid balls—a model clearly drawn from everyday experience. Then, as physicists discovered that atoms were composed of more elementary, subatomic particles (electrons, protons, neutrons, and suchlike) the model shifted to one of a central nucleus surrounded by orbiting electrons—again, a model based on experience.

An atom may be small, a mere billionth of an inch across, but subatomic particles are a hundred thousand times smaller still. Imagine the nucleus of an atom magnified to the size of a golf ball. The whole atom would then be the size of a football stadium, and the electrons would be like peas flying round the stands. As the early twentieth century British physicist Sir Arthur Eddington put it, "Matter is mostly ghostly empty space." To be more precise, it is 99.9999999% empty space.

With the development of quantum theory, physicists have found that even subatomic particles are far from solid. In fact, they are nothing like matter as we know it. They cannot be pinned down and measured precisely. Much of the time they seem more like waves than particles. They are like fuzzy clouds of potential existence, with no definite location. Whatever matter is, it has little, if any, substance.

Our notion of matter as a solid substance is, like the color green, a quality appearing in consciousness. It is a model of what is "out there," but as with almost every other model, quite unlike what is actually out there.

Even the notion of mass is questionable. In his General Theory of Relativity, Albert Einstein showed that mass and acceleration are indistinguishable. When we are in an elevator we feel lighter when the elevator accelerates downward, and heavier when it decelerates to a halt. This is no illusion; scales would also show our weight to have changed. What we experience as mass is the resistance of the ground beneath our feet to our otherwise free fall toward the center of the Earth. According to Einstein, we are being continually decelerated, and interpret that as mass. An astronaut in orbit experiences no mass—until, that is, he bumps into the wall of the spacecraft and experiences a temporary deceleration.

Whatever matter is, it is not made of matter.

HANS-PETER DÜRR

SPACE-TIME AND ACTION

Einstein's work also revealed that space and time are not absolutes. They vary according to the motion of the observer. If you are moving rapidly past me, and we both measure the distance and time between two events—a car traveling from one end of a street to another, say—then you will observe the car to have traveled less distance in less time than I observe. Conversely, from your point of view, I am moving rapidly past you, and in your frame of reference I will observe less space and time than you do. Weird? Yes. And almost impossible for us to conceive of. Yet numerous experiments have shown it to be true. It is our commonsense notions of space and time that are wrong. Once again they are constructs in the mind, and do not perfectly model what is out there.

Kant foresaw this a hundred years before Einstein. He concluded that space and time are the dimensional framework in which the mind constructs its experience. They are built into the perceiving process, and we cannot but think in terms of space and time. But they are not aspects of the objective reality. That reality, according to Einstein, is something else, what he called "space-time." When observed, space-time appears as a certain amount of space and a certain amount of time. But how much is perceived as space and how much is perceived as time is not fixed; they depend upon the motion of the observer.

If space, time, and matter have no absolute objective status, what about energy? Physicists have a hard time saying exactly what energy is. It is defined as the potential to do work, that is, to create change. Energy comes in many different forms: potential energy, kinetic energy, chemical energy, electrical energy, heat energy, radiation energy. But we never measure energy as such, only the changes that we attribute to energy.

Energy is often said to be a fundamental quality of the cosmos. But that too turns out to be a mistake. According the Special Theory of Relativity, energy and mass are interchangeable, related by Einstein's famous equation, $E=mc^2$. Observers traveling at different speeds will differ in their measurements of how much energy an object has.

Quantum theory offers further clues as to the nature of energy. The quantum is commonly called a quantum of energy, the smallest possible unit of energy. But that is not strictly correct. The quantum is actually a quantum of action.

What is action? It is another physical quantity like distance, velocity, momentum, force, and others that we meet in physics, but it is not usually given much attention in our basic math or physics classes. Action is defined as an object's momentum multiplied by the distance it travels, or the object's energy multiplied by the time it is traveling—the two are equivalent. The amount of action in a ball thrown across a football field, for example, would be greater than the same ball thrown half the distance. Double the ball's mass, and you double the action. Or imagine yourself running with a constant rate of energy output. If you run for twice as long, there will be twice the action—which makes sense intuitively.

The amount of action in a quantum is exceedingly small—about 0.00 0000000000000000000000000662618 erg.secs (or 6.62618×10^{-27} erg.secs in mathematical shorthand)—but it is *always* exactly the same amount. It is one of the few absolutes in existence, and more fundamental than space, time, matter, or energy. The zero-point field is not therefore a potential energy field—despite the fact it is often referred to as such. It is a potential quantum field, a field of potential action.

A photon is a single quantum of light, but the energy associated with a photon varies enormously. A gamma-ray photon, for example, packs trillions of times more energy than a radio-wave photon. But each and every photon, each and every quantum, is an identical unit of action.

When the photon is absorbed—by the retina of the eye, say—it manifests as a certain amount of energy, measured by the amount of change it is capable of creating. This change is what is conveyed to the brain and then interpreted as color. The amount of change, or energy, is dependent upon the frequency, which is why we say different colors correspond to different frequencies of light.

What is frequency? Again it is another model taken from experience and then imagined to apply to the photon. It is most unlikely that a photon has frequency as we think of it. Indeed, even the idea of a photon is another example of how we have projected our experience on to the external world. We experience particles, so imagine light might be a particle. We also have the experience of waves, so imagine light as a wave. Sometimes light seems to fit one description, other times another. It is much more likely that light is neither wave nor particle. For reasons of space, I will not go into the details of the argument here, but the interested reader can find more in my book *From Science to God.*

To summarize the argument so far: Our whole experience is a construction in the mind, a form appearing in consciousness. These mental forms are composed not of physical substance but of "mindstuff." We imagine that the world out there is like the forms that appear in consciousness, but it turns out that, in nearly every aspect, the external is not at all like the images created in the mind. What appear to us as fundamental dimensions and attributes of the physical world—space, time, matter, and energy—are but the fundamental dimensions and attributes of the forms appearing in consciousness.

Matter is derived from mind, not mind from matter.

THE TIBETAN BOOK OF THE GREAT LIBERATION

TWO ASPECTS OR ONE?

In chapter 4 of this book, Ervin Laszlo introduces *panpsychism:* the hypothesis that consciousness is not unique to human beings, or higher animals, or even creatures with nervous systems. It is in everything. As he is at pains to point out, this is not to imply that simpler systems have thoughts or feelings, or any of the other mental functions that we associate with consciousness, only that the capacity for consciousness is there in some form, however faint. Even a lowly bacterium has a glimmer of the inner light, maybe a billionth of the inner light we know, but not nothing at all.

The standard scientific paradigm assumes the exact opposite—that matter itself is completely insentient, completely devoid of the capacity for experience. Consciousness only comes into existence with the evolution of complex nervous systems. The problem with this view—David Chalmers' "hard problem"—is explaining how conscious experience could ever emerge from insentient matter. Why doesn't all that neural processing go on "in the dark?"

Laszlo argues that the only tenable answer—anathema as it may be to the standard scientific worldview—is that the capacity for inner experience does not suddenly appear, as if by magic, once a particular level of complexity has arisen. The potential for inner experience has been there all along.

Panpsychism is usually taken to imply that there are dual aspects to everything. There is the physical aspect, that which we can observe from the outside, and there is a mental aspect, the experiences known from the inside. For a long time I went along with this "dual aspect" view. But recently I have begun to question it. I have not questioned whether or not there is a mental aspect, which is the question that most people raise. I have come to question whether there is, after all, a physical aspect. I realize this is radical to many, but let me briefly go over the reasons behind this suggestion and the implications.

Every time we try to pin down the physical aspect we come away empty-handed. Every idea we have had of the physical has proven to be wrong, and the notion of materiality seems to be evaporating before our eyes. But our belief in the material world is so deeply engrained—and so powerfully reinforced by our experience—that we cling to our assumption that there must be some physical essence. Like the medieval astronomers who never questioned their assumption that the Earth was the center of the universe, we never question our assumption that the external world is physical in nature. Indeed, it was quite startling to me when I realized that the answer might be staring us straight in the face. Maybe there really is nothing there. No "thing" that is. No physical aspect. Maybe there is only a mental aspect to everything.

We would then have to think of the Akashic Field as a field that is entirely mental in nature. Its essence is the essence of mind. It's hard to imagine, I know. In fact all we *can* imagine are the forms arising in our minds. We cannot imagine consciousness itself. It is the imagin*er,* that in which images arise. It is probably best not even to try to imagine what a mental field is like, for we would surely be as wrong as when we try to imagine quanta, or space-time.

All we can say about it is that it is not a uniform field. It must contain distinctions of some kind, for it is these variations that are the origin of our perception of the world. If there were no variations in the field, there would be nothing to observe, nothing to experience.

These variations in the field are the "objects" of our perception. But they are not objects in the sense of a material object. They only become material objects in the mind of the observer. There then appears to be a material "thing" out there. We then assume that the physicality we experience, which seems so intrinsic to the world we know, must also be an intrinsic aspect of the external world.

Even though there may be no physical basis to the external world, the laws of physics still hold true. The only thing that changes is our assumption about what we are measuring. We are not measuring physical particles or such, but perturbations in the Akashic mind-field. The laws of "physics" become the laws governing the unfolding of a mental field, reflections of how perturbations in this field interact.

What we call an elementary particle thus corresponds to an elementary variation in the field. We might better call it an elementary "entity" rather than particle. Elementary entities are organized into atoms, molecules, cells, and suchlike, just as in the current paradigm. The difference is that we no longer have to think of consciousness sensing matter (with all the difficulties that involves regarding how the physical influences the mental); consciousness is actually sensing consciousness directly.

Interaction can thus be thought of as perception—the perception of one region in the mind-field by another. In the prevailing view every interaction is mediated by a quantum of action (an inter-action). In this alternative view, the smallest item is a unit of perception, a unit of experience. It is a quantum of consciousness, a quantum of chitta.

In the physical world of our experience we have discovered action to be a fundamental quality. In this alternative view, that still is true.

Consciousness acts as it takes form. A quantum of action is a quantum of experience, a quantum of chitta.

We can now begin to understand why the material world appears to be devoid of consciousness. The qualities that appear in the mind—the color, sound, smell, substance, or whatever—become objects of perception, "the material world." But there is no sign of consciousness itself in the images of matter that appear in the mind. It is similar to our experience of watching a movie: the picture on the screen may be composed of light, but there is no evidence in the unfolding story that this is the case. The forms that arise in the mind give no hint in themselves that they are all manifestations of mindstuff. They appear to be other than consciousness. And so we assume that the stuff of the world "out there"—the matterstuff—is insentient.

Physics is the study of the structure of consciousness.
The "stuff" of the world is mindstuff.

Sir Arthur Eddington

THE HARD QUESTION REVISITED

The hard question of how insentient matter could ever give rise to conscious experience is now turned inside out. There is no insentient matter—apart from that appearing in the mind. The question now becomes: How does mind take on all these qualities that we experience, including that of matter?

That question is best answered by direct awareness, by turning the light of consciousness in upon itself, and observing the nature of mind first-hand. Those who have chosen this path are the great mystics, yogis, seers, saints, rishis, and roshis who are found throughout human history.

Despite the differences in time and culture, they have come to remarkably similar conclusions. These conclusions do not, however, make much sense to the contemporary Western mind. In most cases they seem to be so at odds with the standard scientific worldview that they are rejected out of hand—and with them any credibility there may be for spirituality in general.

Consider, for instance, the statement by Baba Muktananda that "You are the entire universe. You are in all, and all is in you. Sun, moon, and stars revolve within you." Most people would be puzzled, if not confused. It clearly goes against the contemporary worldview in which I am a small point at the center of my universe, around which everything else revolves. Muktananda appears to be saying the exact opposite. Possibly, we might surmise, a mind deranged by too much meditation.

However, if we see it in terms of an intimate personal acquaintance with the arising of mental phenomena, and hence of our whole world, it makes much more sense. Every experience, every thing we ever know, is taking place within us.

Likewise, when we read such peoples' accounts of creation, we are likely to interpret them in terms of how the physical world was created. In a sense they are. But they are talking of the physical world as it appears in the mind—how it is being continually created.

The *Ashtavakra Gita,* a highly venerated Indian text, says: "The Universe produced phenomenally in me, is pervaded by me. . . . From me the world is born, in me it exists, in me it dissolves." Hardly comprehensible, until we consider it from the point of view of consciousness.

"In the beginning was *Logos.*" Often translated as "The Word," logos also means "thought, or essence." In the beginning was the mental essence, chitta.

"Be still and know that I am God" is not necessarily an injunction to stop moving around and recognize that the person speaking is the creator of the entire cosmos; it is much more likely an encouragement to still the mind—in the words of the great yogi Patanjali, "let the manifesting of chitta die down"—and discover through direct knowing, that "I", that ever-present, never-changing, innermost essence of one's own mind, is the essence of everything.

This view is echoed in an earlier book by Ervin Laszlo, *The Connectivity Hypothesis.*

> The evolutionary process occurs within God's consciousness. It leads from an initially unruffled divine consciousness informed by pure potentials, through the spatiotemporal excitation of this consciousness, to a finally unruffled consciousness in-formed with the full record of the realization of those potentials.

In this I find a personal reenchantment of the cosmos. If our own essence is divine, and the essence of consciousness is to be found in everything, everywhere, then everything is divine. Panpsychism becomes pantheism. It doesn't matter whether we call it Universal Mind, Allah, God, Jehovah, the Great Spirit, or the Quantum Vacuum Field, we are all of that same essence.

This raises my level of awe for the world in which I live, or seem to live. When I consider that—despite all appearances to the contrary—this world is, in the final analysis, of the same essence as my own being, I am filled with wonder. Every thing is enchanted anew.

The Reenchanted Cosmos and the Sufi View of the World

Johannes Witteveen

In this book Ervin Laszlo makes a further contribution to what is one of the most profound and fascinating developments in present-day thinking: the reconciliation of science and spirituality. These two fundamental forces have diverged disastrously since the Renaissance. The traditional image of God put forth by the monotheistic religions has become untenable in the light of scientific insight. Many people have expressed this divergence with the terrible saying, "God is dead." What is left is an awful and painful emptiness at the heart of our culture. Now, with the recent discoveries made by scientists, science and spirituality are once again approaching a more harmonious relationship.

Laszlo follows a large number of scientists working in various fields as they attempt the difficult ascent upon the mountain of truth. He shows that from their different vantage points they discover glimpses of a truth that goes far beyond the traditional physical model of the world, confirming in a striking way what mystics through the ages have found and experienced intuitively. He is happy to point this out, quoting Hindu and other mystical thinkers and even animist cultures. But we need not go back that far in time. The philosophy of Hazrat Inayat Khan, the great Indian mystic and musician who presented universal Sufism as an answer to the fundamental questions of our time, is exactly in line with the scientific vision that Laszlo—thanks to his wide knowledge of the new findings—has produced.

The mystical expression of truth is more poetic and appealing to our deepest being than abstractly formulated scientific conclusions. But the convergence of the two is very important for our culture. It validates the mystical vision for our thinking mind, which gives our hearts permission to be inspired. A present-day spiritual worldview that satisfies our deep human longing can restore meaning to our lives and rekindle our values. A mystical poetic expression of truth can bring the "enchantment of the universe" to life in our hearts and in our culture. Let me describe here a few instances of this convergence.

A crucial concept of the new scientific vision is *coherence*. Scientists have found everything in our bodies, in our relationships, and in the universe, to be immediately connected over any distance in "non-local relationships." In living organisms this works with amazing sensitivity. Laszlo quotes the biophysicist Mae-Wan Ho, who compares the living organism with a good jazz band in which every musician responds immediately and spontaneously to the improvisations of all the others. Inayat Khan, in turn, uses the word *harmony* and writes in his work *The Mysticism of Sound and Music*, "Life is the outcome of harmony. At the back of the whole creation is harmony, and the whole secret of creation is harmony." He continues, ". . . man, being a miniature of the universe, shows harmonious and inharmonious chords in his pulsation, in the beat of his heart, in his vibration, rhythm and tone. His health or illness, his joy or discomfort, all show the music or lack of music in his life."

Coherence exists and works in the world because there is a medium that connects all things and beings. In the past this was called the "ether" *(Akasha)*. Modern science sees it as the quantum vacuum, an unending field of virtual energy. This fundamental field is not empty; it actually proves to be full of energy. The particles that form all material objects and beings constantly arise from this field. Periodically, they also dissolve back into it. This connecting medium keeps in contact with all of the diverse elements of the physical universe. Because they absorb the energy they need from this field, it also maintains their stability.

Laszlo shows a further important aspect of the quantum vacuum: it preserves information. All actions, thoughts, and feelings, all experiences in the universe are registered in it. These qualities of the quantum vacuum correspond precisely to the mystical vision described by Inayat Khan. In philosophical language he made a distinction between spirit and matter. "Spirit is the absence of matter, even in its finest condition." In physical terms, he compared spirit and matter to water and ice. "Just as ice and water are two things and yet in their real nature they are one, so it is with spirit and matter. Water turns into ice for a certain time, and when this ice is melted it will again turn into water. Thus matter is a passing state of the spirit; only it does not melt immediately as ice melts into water, and therefore man doubts if matter, which takes a thousand forms, ever really turns into spirit. In reality matter comes from spirit; matter in its true nature is spirit; matter is an action of spirit which has materialized and has become intelligible to our senses of perception, and has thus become a reality to our senses, hiding the spirit under it."

In this way Inayat Khan saw all things coming from spirit and returning to it. He saw the quantum vacuum, or the Akashic Field, as spirit, and this spirit as the Divine Spirit or Divine Being. This is the Only Being, All-pervading, All-powerful, the creating and sustaining Spirit (in scientific terms: the field that maintains stability), and the All-knowing Spirit (in scientific terms: the field that preserves all information), that which registers the experience of all beings in the divine mind or divine heart. As he puts it, "In this mechanism of the world, all that happens, all that is experienced in the way of thought and feeling, is accumulated. Where? In the heart of God. . . . the heart of man is one of the atoms which form the heart of God."

The scientific and the mystical views of the world are converging in other ways as well. The statistical probability that our universe would have emerged as a consequence of purely random selection among myriad possibilities has turned out to be inconceivably small. It appears that there must be some direction to the creative process, which points to a purposeful creative spirit. But Laszlo mentions another possibility envisaged by scientists. Our universe could be one of a multitude of universes successively created and destroyed. Thus, over endless periods of time, the information needed to produce our amazingly coherent universe could have been gradually built up. The propensity of our universe to bring forth life would then be a matter of heritage.

The Sufi vision of creation comes close to this understanding. In this mystical vision, God is not a separate higher being but the all-pervading spirit that works in all elements, all atoms, plants, and beings of the cosmos, giving them energy, light, and direction. The progressive creation and destruction of universes is seen as the breathing of God. Inayat Khan wrote: "Creation is the exhalation of God, and what is called destruction is absorption, which is the inhalation of God. . . . The beginning and the end of the world is only His one Breath, the duration of which is numberless years. During this one Breath myriads of beings have been born, have lived and died and experienced this world and the next."

This scientific-spiritual worldview is of great importance to our daily lives, informing the choices we have to make continuously. It is the human privilege and responsibility to make conscious choices in life. But what determines which choices are right and which are wrong? In the context of the new worldview, Laszlo notes the determinant influence of our thinking and acting on the co-evolutionary process. Right and wrong depend upon *coherence*. Our choices can have a positive or negative influence on the functioning of the coherence that is in and around us. Within us, good coherence means good health—the optimal functioning of our bodies. Around us, it means the optimal functioning of society and the natural environment. In the Sufi vision this ideal of harmony, which requires love and creates beauty, makes us happy and gives us the inner experience of the enchanted universe.

The Oneness of Objective Reality and Spiritual Truth

Swami Kriyananda

I feel honored to make a contribution toward this valuable book, the subject of which is close to my own heart. I have tried all my life to convince people, as Ervin Laszlo has done from the scientific perspective, that, contrary to the insistence of traditional scientists, there is deep meaning in life, and that materialism—no less so than most religious orthodoxy—is (as he puts it) "sophistry," or, as I sometimes put it, "superstition."

There is a growing awareness today—and it is exciting to contemplate—that science and spirituality are actually motivated by the same aims: the desire to know *what IS*. Among what may be called researchers in both fields, there is a growing sense that a unifying, conscious reality exists. The meeting ground between them lies in the ancient Vedantic dictum: "Truth is center everywhere, circumference nowhere." The aim of spiritual research is to withdraw to the center of one's being: at the heart of one's own energy and consciousness, and there to discover one's Self as the heart of all reality! From one's own center, so great saints and yogis have found, it is possible to reach out and understand the meaning of existence itself. That truth lies at the center of *every atom*.

From what Laszlo calls the cutting edge of modern science, matter itself is beginning to look like a manifestation of one central, all-unifying reality. It is thrilling to find that these two lines of research—apparently and for so long disparate—not only run along parallel lines but, by some strange paradox, are making identical discoveries.

It is axiomatic to any search for understanding that one must begin with what one already knows. Science has done that from the beginning, a fact that explains why scientists have for so long rejected the fallacy that certain beliefs in religion must be accepted blindly. The same axiom explains also why science now finds itself face-to-face with realities that are far subtler than the material.

How much do we actually know, living as we do through our five senses? Do we really *know* that the sun rises? That the moon waxes and wanes? That we are alive, and that life inevitably ceases in what we call death? None of these things can be known, by ordinary perception, with any certainty. The sun doesn't rise: it only seems to do so as the earth spins on its axis. The moon doesn't wax or wane: it only seems to do so according to its changing positions relative to the sun and the earth. As regards life, we know only that we *experience* what appears to be life, and that, eventually, our experience ends. All things perceived through the senses are uncertain. We live in a world of appearances. Colors, solids, shapes—these and countless other phenomena have no basic reality: They are, all of them, mere seemings.

If the nature of our research is spiritual, the outward uncertainties I have described inspire us to look inward to our own center. What are we, really? Do we even exist? A further doubt arises: How different is objective reality from what we see when we dream at night? From thoughts like these, serious spiritual research begins.

René Descartes pondered the problem of existence, and finally reached the conclusion: "I think, therefore I am." His dictum has been acclaimed for centuries as one of the great discoveries of human thought. And yet, careful analysis shows that even this statement lacks a firm basis. It is possible *not* to think at all, and *still* be aware. It is *awareness*, not thought, which confirms our existence. Indeed, one can be much more aware *when not thinking at all:* as, for instance, in a deep, meditative state. Descartes' reasoning was brilliant, but it had all the rainbow beauty of a bubble.

There is a deeper knowing, available to us all, which depends not on any web of reasoning—so often fragile!—but on the insights of irrefragable intuition. We have no need to convince ourselves logically that we exist, for our *intuition of the fact precedes thought itself*. Even the humble earthworm, which (one feels safe in assuming) does not think at all, is aware enough of its existence to wriggle away from anything that it perceives as a threat.

The awareness we have that we exist may be termed *intuitive*, inasmuch as it depends not on objective testimony, nor on any informed reasoning: It simply IS. Only from that initial, intuitive insight of our existence can we proceed, spiritually, to investigate the meaning and purpose of life.

Ervin Laszlo writes: "Science at its cutting edge rejects the idea of a material universe heading ineluctably toward a desolate destiny. The new cosmology discovers a world where the universe does not end in ruin, and the new physics, the new biology, and the new consciousness research recognize that we are part of an instantly and enduringly interconnected integral reality. A reality that is connected, coherent, and whole recalls an ancient notion that was present in the tradition of every civilization; it is an enchanted cosmos." These thoughts are thrilling. Surely the long-standing opposition—has it not been, rather, a feud?—between science and religion will soon be resolved by the discovery, on the one hand, that objective reality is only a projection of consciousness through the medium of energy; and, on the other hand, by the recognition among spiritual seekers that the search for meaning is indeed a quest, and therefore does not demand a blind acceptance of orthodox dogmas. Religion must be transformed into spirituality—that is to say, into a search for the meaning of existence itself, at the heart of all reality. And science, as Laszlo has shown, is discovering the same truth from outside, by its investigation into the objective nature of reality.

Indeed, objective reality and spiritual truth are one!

Reconnecting With Nature and Cosmos

Irene van Lippe-Biesterfeld

In the twenty-first century we find ourselves within a significant transformation, in a shift from a mechanistic and segmented way of thinking to an all-encompassing integrated worldview. We are revaluating our position in nature and in the cosmos. The energy of this transformation is like a great stream flowing through all that is, touching each individual, each life form. It is happening whether we know it or not.

In this transformation we can step into freedom, into the joy that living together with all the different cultures and life forms can give us—into the "we" proposed by Thomas Berry. As the anthropocentric way of life has proven to be a dead-end, many people are making this shift. We are coming at last to understand that we are facing the fundamental choice: to keep on devouring the Earth's resources, or being open to an integrated worldview. Our decisions and the consequences of those decisions will create a totally different reality for the planet.

Through my own personal path I have come to understand intimately that we are part of the Great Community of Life. We are both biologically and spiritually part of the great community and therefore need to reconsider our role in this oneness. All non-human forms of life are part of this community, but they do not have the rational way of thinking that comes in between the Source they carry within and the

Source without, of which they are a part. Our quest for finding our inner nature and allowing ourselves to work and live from a whole—and holy—perspective, is imperative.

To me it is clear that if we cannot accept ourselves we will not live in peace. What is inner peace? To me it is what we can achieve through the acceptance of who we are, by getting to know our essence or inner nature and coming to understand why we act as we do in the various circumstances and contexts of our life. And by recognizing that we belong to the beautiful truth that we carry within ourselves as an interconnected part of the whole. Ours is a biological connection as well as a spiritual one: it is "interbeing" as the Vietnamese Buddhist monk Thich Nhat Hanh calls it. Through this connection we can discern what love really is. Clarity is an important aspect of love and the ability to put the ego aside. Through learning to look at and overcome our fears, we can grow step by step into love, respecting and accepting *ourselves,* totally together with *all that is.* The deeper we embrace our inner nature, the deeper we can connect to our surroundings. Every living thing is there for its intrinsic value, for serving the others, as well as for the whole. It is the same for us humans.

Our disconnection from the Earth and all that lives on and above the Earth has been a huge and horrible mistake that has led us to destroy a great part of the planet and its inhabitants, destroying at the same time parts of ourselves. What this has done to the cosmos, we can only guess. This is why inner healing and the healing of the Earth go together. Our role as a species carries a deep responsibility, because we can be conscious of what we do.

The change of mentality leading to interbeing may appear radical, as it entails a huge mind-shift. We need to dissolve the barrier between the person we think we should be in our working situation, and the person we are in our home situation. Whole people can make whole decisions. The mind-shift requires a connection between personal growth and economic growth. We cannot do business, teach, and help others without involving ourselves as we really are. Only through the growth of inner freedom can things change. Openness is an essential part of this growth, even when it demands that we change our view of ourselves and of life.

We *can* feel the oneness of all life, we *can* experience it, as many have testified, and this can change one's life—it has changed mine. Every change we make, every fear we conquer and transform into love, resonates with that embracing community of life of which we are a part. Ervin Laszlo explains this and gives it a sound basis. The Akashic Field registers all we think of and intend, as it registers everything in nature. Understanding the Akashic Field brings what we have known in the world of healing from prehistoric times to a scientific level. In the sphere where inner growth and consciousness was the focus, the Akashic Field was seen as a field filled with information, like a library where we can tune into and receive information unlimited in time and space. Knowledge about this Field was based on mystic experience and philosophical insight. It comprised all the information about all that is and ever was on this Earth and in the universe.

The traditional cultures still live in oneness and interbeing; they are actively connected to all life, and all life is connected to their inner nature in their daily life. Nature is embedded in traditional culture as a deep source of information. I am not saying that we should live like traditional people. But the more we become multisensory persons, as Gary Zukav wrote, the better we can relate, feel, and experience the interconnectedness of all things, and open up to mutuality in our relationship with the many forms of life around us. A multisensory person can feel the energies not only of other people, but also of the flowers, the sun, the trees, the Earth's energy, and the coming of natural catastrophes like earthquakes or tsunamis. We carry this ability in us, but we have not used it for some time; a little exercising will revitalize our faculties.

We need to start with *wanting* to be open and receptive; to be open vis-à-vis other cultures and life forms. We need to want to emerge from the rat race of contemporary society, to stop from time to time and make space to feel and to listen. To find the strength that lies within us, and from there look at what needs to be done in the outside world. To allow a change to take place in perception so as to look at others from soul to soul; to look at all forms of life from soul to soul.

How you treat someone and make a decision, how you feel when you arrive home, all this is felt even by your pet and by the trees and greenery around you. You need to feel that you are part of other people, of nature, of the cosmos, to become a conscious part of the magnificent all-encompassing biosphere. The moment you are in contact with your uniqueness, your personal dynamics, you are no longer an obscure presence. On the contrary: your footstep, your smile, your intelligence, your insight, your eyes, give the world something totally unique.

Yes, we should open up to all the different life forms around us. We are not alone and it is up to us to be open and receptive to the oneness, the continuing interconnection that can make us strong, conscious, and more than the mere sum of a mechanical system.

Should we not and can we not see that we have to change our mindset toward saving all life for all life? Our survival instinct should prevail; we need to become more and more aware of the havoc we have caused around and in ourselves, and we must choose to preserve the planet for our grandchildren, and the grandchildren of the ants and the birds, the grasses and the whales! We have taken so much space that many animals now are only allowed to live in zoos and in restricted nature parks.

It is up to every one of us to make the inner decision: to continue to live with the illusion of separateness, or live with the awareness of the oneness of all life in the beautiful Symphony of Life. Reconnecting with life brings the song back into our hearts. Being open to the animals and elements of nature makes our life rich in creativity. It makes us embedded in the great community of life.

The entire Earth with the animals and all her creatures as well as the entire cosmos is the context for the urgently required shift of mentality from separation to oneness. All of us together can work in extraordinary and amazingly interconnected ways to rediscover and repair our interconnections and interrelatedness. Let us not make this choice from fear or from lack, but with the clear intention of preserving the infinite beauty of the many forms of life on this planet. The same as the reenchanted cosmos itself, reconnecting with it is also a happy re-discovery!

A Mystic Journey into the Reenchanted Cosmos

Ewert Cousins

Science and the Reenchantment of the Cosmos is an extraordinary book that further illumines the remarkable work Ervin Laszlo has already produced. Over many years I have caught glimpses of the universe he has seen with such clarity. During the last several months, as I read first *Science and the Akashic Field* and now the *Reenchantment of the Cosmos,* I have felt my view of the universe expand in a way I never dreamed possible. I would like to give three examples from my own experience of the Akashic Field as "an information field that pervades the universe."

A New Vision of Consciousness from Indigenous People. The first vision is from my personal experience of contact with indigenous people: with the Pottawatomie Indians (the People of the Fire) in Kansas from 1954 to 1958 and with the Lakota or Sioux in South Dakota in 1957 and 1958.

In both cases I had extensive personal contact with them in such a way that my own experience was interwoven with theirs. My general perception was that they lived in "two realms of consciousness," one that can be called "the white man's world" and their own realm of indigenous consciousness. Because of this dual mode and the richness of their "indigenous" consciousness, they are among the most sophisticated people I have ever met.

One of the most striking phenomena that I observed was their instant communication among themselves across great distances. For example, if we announced at our little mission church on Sunday that several Pottawatomie children miles away at the other side of the reservation were in urgent need of a home because of a crisis, instantly this would be known at a great distance across the reservation and a family would come forward that wished to take them in and care for them as one of their own. This kind of instant communication was common. Non-natives at the church used to marvel at this phenomenon. There are many examples of this instant communication among indigenous people. I now see it as an example of the function of the Akashic Field.

This was true not only of the Pottawatomie but also of the Sioux, who were much closer to their traditional way of life than the Pottawatomie. During my two summers on the Sioux reservation, I lived and worked very closely with the Indians, branding cattle and repairing and painting their school. In the evenings after work I would often saddle up a horse and ride into the canyons to visit Chief Hollow Horn Bear, the last official chief of the Brulé Lakota. When I asked him to describe the traditional Lakota religion, he told me the story of the White Buffalo Woman, a heavenly figure who miraculously appeared to two Lakota braves on the plains. She prophesied that an animal that they had never seen before would come to their village. If the Lakota welcomed and honored it, then the animal would greatly enhance their lives. This was reported to the tribe who waited in expectation.

A short while later a herd of buffalo ran through their village. When the dust had settled and the buffalo had departed, the Lakota discovered a strange animal that did not resemble a buffalo; rather it looked like a large dog, so they named it *chunckaka* or "big dog." To this day, the Lakota call it "big dog." This animal was the horse! Soon other horses came, and with them the Lakota became the most powerful tribe throughout the plains. On the "winter count," an annual chronology of tribal events, the first time the horse appears is the year Coronado crossed the plains and brought with him horses from Spain. As the White Buffalo Woman had foretold, the gift of the horse enormously enriched the future of the Lakota Nation.

In this atmosphere, I found myself penetrating deeper and deeper into Lakota culture, their dramatic history and their profound spirituality. I vividly remember the day, while I was talking to a group of Lakota, that I felt my consciousness, as it were, extend itself out of my body and pass over into their consciousness. From that moment I felt I could see things from their perspective and experience their values from within their world. Also I could look back at my own world and see its values in a clearer light—but also its limitations! In retrospect now I realized that my consciousness had entered into the Akashic Field.

When I returned to Kansas, my Pottawatomie friends had detected that I had resonated in a new way with the Indians. They suggested that I might be interested in meeting with the leader of the ancient Pottawatomie religion, called the People of the Drum. I met their leader, Curtis Piquano, at a naming ceremony for a young boy that included lengthy beating of drums and ceremonial chanting near a stately tepee set up for the occasion. We talked together for three hours. I can say sincerely that Curtis Piquano was one of the wisest, most articulate, and profoundly spiritual persons I have ever met. He described their ancient traditions—the meaning of the drum, their ceremonial gathering four times a year out of doors, even in the dead of winter, with the beating of drums and chanting for hours. He explained their vision of the sacred pipe and the sense of the sacred that penetrated their lives. On a sad note, he told me how the Smithsonian Museum often contacted him for permission to record the chants of the Pottawatomie but he felt that he should refuse because he said their traditions were dying and that the chants should die with them. Perhaps he saw that their chants would be preserved in the Akashic Field.

The Akashic Field and Psychological Research. The basis for my exploration into the psyche has been my collaboration in the research of Jean Houston and Robert Masters. From the mid-sixties to the late seventies I participated in their research in states of human consciousness.

This research has revealed that the psyche has four levels, which are described as follows: (1) the sensorium, or the level of heightened sense experience; (2) the ontogenetic, or recollective analytic level, where the subject recalls his or her personal history, which corresponds to Freudian psychoanalysis; (3) the phylogenetic or symbolic level, corresponding to the collective unconscious of Jung, in which the subject relives the great myths and rituals of humankind; (4) the level of integral consciousness, or the mysterium, which is the level of deep mystical consciousness, similar to that described by the mystics of the world.

These levels are markedly different, with their own horizons, structure, logic, and dynamics. Although there is a certain interpenetration of each in the others, they remain sufficiently distinct; yet one leads to the other.

In my own experience, subjects usually experienced heightened sensation at the beginning of the process of entering an altered state of consciousness. For example, having been handed an orange by the guide, a subject contemplated it for several minutes and said: "Magnificent . . . I never really saw color before . . . It's brighter than a thousand suns . . . (Feels the whole surface of the orange with palms and fingertips) But this is a pulsing thing . . . a living pulsing thing . . . And all these years I've just taken it for granted."

After some time on the sensory level, subjects move to the second level, that is, the ontogenetic or recollective analytic, and then to the phylogenetic or symbolic level. I will now give examples of my own experience as a subject, first of the ontogenetic level and then of the phylogenetic. When I was not quite four years old, I had experienced deep emotions at the time of the death of my grandmother. In my session I suddenly said to myself, "I feel as if some obstacle is coming up—something large, dark, and vague, but very powerful—as if it were knocking on the walls of consciousness . . . It's Granny's death! I must examine Granny's death!" I then felt a surge of guilt that I had experienced years before over my grandmother's death, but which I had repressed. In the course of the session I was able to free myself from the burden of this unconscious guilt.

My experience on the phylogenetic level was simply remarkable! I entered into a realm of consciousness I had never visited before nor have I since. For a period of about four hours, I found myself with a group of indigenous people going through an elaborate puberty rite of induction into the tribe. Through the years I eventually went on to explore other realms of consciousness, but none could compare with this. What I describe below gives only a surface glimpse of what I experienced. In retrospect it seemed that I had entered the Akashic Field where I would remain for four hours. I was in a different dimension of consciousness—in a different time and space.

It began with experiences, image sequences consisting of a series of rites. These were:

- A fertility rite. This consisted mostly of white-skinned savages dancing and singing around a fire, "trying to bring something to life." I was both emotionally and through bodily sensations a participant in this rite, but my involvement was not great and the rite lasted only a few minutes. It seemed to me to be a preliminary to something more important still to come.
- A warrior initiation rite. This rite also began with dancing, which became increasingly faster and more violent. Then the young candidates for initiation liberated themselves from paternal influence and so became ready to assume their roles as warriors and leaders of the tribe. I also was fully involved in this rite.
- Then I had the awareness that I was lying down on my back and that someone was placing hot coals in a circle on the lower part of my abdomen. I was afraid. Then I accepted the situation and entered into the ritual. At that moment, a man appeared in front of me. I knew that he was the Savior. I could not discern his features. His face seemed to be white, without any features, and I could see only his bust. As soon as he appeared, I threw over my left shoulder a piece of animal skin; it seemed to have hair and to be a piece of goatskin.
- Then I noticed that the people were on a field and were tearing the Savior to bits and eating his flesh. Then I felt that I was the Savior

> and was lying on my back being nailed to a cross. Then the cross was lifted up, and at this moment I was a spectator viewing the Savior from a distance as he was being lifted up for all the people. From the time the Savior appeared, I had a deep sense of peace and integration. I felt that I was saved and that I was whole. This intense drama lasted for about four hours.

I was then exhausted and slept briefly. I awakened and thereafter described myself as very tranquil and happy. I remained silent for some time and then reported that I had been "integrating" all that had occurred, and that this integration was extremely important. I felt drastically and beneficially transformed. I viewed my image in the mirror and said it was greatly changed. My face now was "much more youthful and relaxed" than it had been for years. I felt "strongly related" to everything around me. My appetite, when a meal was served, proved voracious, and I said that I had never felt better or enjoyed food more.

This was an overwhelming experience—unique in my life. I felt I had entered another realm of reality while the drama continued for more than four hours. I deeply felt that I was a member of a prehistoric tribe and was actually going through a puberty rite as a powerful transformation into this tribe that had lived perhaps five thousand years ago, very likely in prehistoric Europe. Even though I recognized a Christ figure, I was aware that in such altered states our ordinary sense of history is altered. I had never had such a sense of being in another time. In view of my present understanding, I would say that I had been drawn into the Akashic Field and accessed the experiences of a prehistoric tribe and participated in their rituals.

On a number of occasions, as I continued to work with Jean and Bob, I had the overwhelming experience of entering the fourth level of the psyche which they called the mysterium.

The Reenchanted Cosmos and the Spiritual Masters. Many of the elements in the "reenchanted cosmos" have a counterpart in the writings of the spiritual masters. The void or dark vacuum out of which comes the plenum is amply illustrated in the spiritual writings of

Saint Bonaventure, who followed the early church fathers in enfolding much of the Platonic tradition into his wisdom. Bonaventure speaks frequently of the *fontalis plenitudo* or the "fountain fullness" in the Trinity and the fecundity of the Father. But this fecundity comes from a void—a privation or negative aspect of the Father, who comes from no one, no thing, is unbegotten and without origin. From this privation or nothingness comes the primacy of the Father and the power and the plenitude flowing in the Trinity.

Boundless creative power is potentially present in these spiritual writings. The Father endows the Son with all of his own power, wisdom, and goodness and yet is not himself deprived. Thus, the Son contains all the eternal ideas in the Divine Mind, a Platonic concept. In him is all that has been created, all that is created, all that will be created, and all that has the potentiality to be created: all that was, is, will be, and may or may not be but has *potentia*.

Bonaventure would have categorically rejected the "mechanistic, materialistic, and reductionist image" of the cosmos. Speaking of the glories of creation, he exclaims that "Whoever . . . is not enlightened by such splendor of created things is blind; whoever is not awakened by such outcries is deaf; whoever does not praise God because of all these effects is dumb; whoever does not discover the First Principle from such clear signs is a fool."

In describing creation, Bonaventure "proclaims the divine power that produces all things from nothing": their origin, magnitude, multitude, beauty, fullness, activity, and order. These properties, claims Bonaventure, "clearly manifest the immensity of the power, wisdom, and goodness of the triune God, who by his power, presence, and essence exists uncircumscribed in all things."

Moreover, the divine presence is evident in the divine vestiges, literally "footprints" in the sense world, or in more modern terms, that part of the cosmos that is matter or observable. Here we see beauty, order, prefiguration, and divine realities shining forth.

On a higher level, the human person can contemplate God "through his image stamped upon our natural powers." In the deepest part of the mind, heart, and soul of the human person is hidden the image of the Trinity in memory, understanding, and will, representing the three persons of the Trinity.

Living up to our potential as conscious beings is a repeated theme among the early Fathers and the mystics who followed. Perhaps no one expresses this better than Iraneus, who said, "The Glory of God is the human person fully alive."

A contemporary mystic, Pierre Teilhard de Chardin, who died in 1955, was a philosopher, theologian, paleontologist, and a Jesuit priest. He began his scientific explorations in "matter-matter," solid stones, and only later became interested in energy. Eventually, he came to see matter and energy, or spirit, not as two things, but two states or aspects of one and the same cosmic stuff, "according to whether it was looked at or carried further in the direction in which . . . it is becoming itself or in the direction in which it is disintegrating."

The cosmos of Teilhard is definitely a reenchanted place in which he felt at home. He writes, "I find it difficult to express how much I feel at home in precisely this world of electrons, nuclei, waves, and what a sense of plenitude and comfort it gives me." Teilhard agreed with Saint Angela of Foligno and a multitude of mystics, who said, "God is everywhere." Meister Eckhart turned this around and said, "Outside of God there is nothing." Teilhard went on to say, "Every exhalation that passes through me, envelops me, or captivates me, emanates, without any doubt, from the heart of God; like a subtle and essential energy, it transmits the pulsations of God's will." "Every element of which I am made up is an overflow from God."

Teilhard described the Earth as being made up of the geosphere, or the actual globe; the biosphere, all the life on our planet; and the noosphere, or the field of knowledge or communication surrounding the world. The noosphere can be seen as a possible corollary to the Akashic Field.

A New Vision of the Cosmos in an Ancient Theme: The World Soul. Similarly to Ralph Abraham, as a final part of my dialogue with *The Reenchantment of the Cosmos* I would like to suggest a complementary concept to the Akashic Field: that of the World Soul. In contemporary language, this pillar of the Platonic-Neoplatonic tradition could well be expressed as the Cosmic Soul. As Abraham points out, it appeared in the fourth century B.C.E. in Plato's dialogue *Timaeus*. It became a classical element in the Platonic tradition, featured in the synthesis of Plotinus in the third century C.E., in the Platonists of Chartres in the twelfth century, and again in Renaissance Platonism.

The World Soul is the energy by which all things become blessed, alive, and moving. It enters the body of the universe and gives it life and immortality. Modern Western theology has failed to take in the total richness of the cosmos, which not only contains its scientific components, but also the spiritual. Laszlo's position is in harmony with the great traditions, not only of Western culture, but of global culture.

An Overview of the Death and Rebirth of the World Soul

2500 B.C.E.–2005 C.E.

Ralph H. Abraham

The individual soul is an ageless idea, attested in prehistoric times by the oral traditions of all cultures. But, as far as we know, it enters history in ancient Egypt. In this overview we begin with the individual soul in ancient Egypt, then recount the birth of the world soul in the Pythagorean community of ancient Greece, and trace it through the Western Esoteric Tradition until its demise in Kepler's writings, along with the rise of modern science, around 1600 C.E. Then we tell of the rebirth of the world soul recently, rising from the ashes of, again, modern science. Recent books of Ervin Laszlo are an important part of this revival of the world soul in new clothes, such as the Akashic Field.

THE WORLD SOUL: BIRTH, LIFE, AND DEATH

1. *Ancient Egypt, 2500 B.C.E.* We take seriously the possibility that ancient Egyptian culture began around 10,000 B.C.E. Thus, tentatively, we may regard it as the Ur source for the soul concepts of the Western Esoteric Tradition, including the Greek and the Indian roots. We begin our story of the history (as opposed to the prehistory) of the soul in 2500 B.C.E., with the Great Pyramid of Cheops. As John Anthony West

remarked, "This world is alive in its entirety and infused with divine spirit." As West further noted, spiritual elements or bodies of ancient Egypt include the Ba, the Ka, and several others.

The Ba, or soul, is "the animating principle, the vital or divine spark that vivifies all sentient creatures." The Ka, or double, is "the power that fixes and makes individual the animating spirit that is Ba. . . . If during life on earth, the Ka has degenerated to the point where it has been divested of all virtue, of everything truly human, then it does not reincarnate, and the Ka disperses into the various lower animal and vegetal realms. . . . It may be this understanding that lies behind the curious doctrine of metempsychosis in which the deceased may be reborn as an animal or even a bush or tree."

Recent studies of the pyramid of Cheops, and the pyramid texts, give an idea of the journey of the soul in the reincarnation process. After death, the Pharoah's soul was said to become a star, to join with Orion in the sky. Alexander Badawy determined in 1964 that the two shafts cut 200 feet from the King's chamber to the surface were aimed at the Pole star, and Orion, in the year 2600 B.C.E. And according to Robert Baumol, the two shafts from the Queen's chamber were aimed at Orion and Sirius in 2450 B.C.E. The supposition is that these shafts were to facilitate the journey of the Pharoah's soul after death and internment of his body in the pyramid, to its home in the sky.

2. Pythagoras, Socrates, Plato, 520 to 347 B.C.E. Pythagoras of Samos was born around 570 B.C.E. He traveled and studied in Egypt and Babylon. Initiated into the mysteries of several traditions—Egyptian, Babylonian, and Persian—he returned to Greece and Magna Graecia in southern Italy and carried on with the reforms set in place by the Orphic religion, which became the most important religion of ancient Greece. Pythagoras synthesized spiritual and natural philosophy into the framework for classical Greek culture, including the metaphysical and sacred aspect of Number, the One (monad, unity) and its emanations. He introduced the terms "philosophy" and "cosmos." He created a school around 520 B.C.E. in Croton (southern Italy) that emphasized communal living, gender equality, vegetarianism, mystery initiations, Orphic poetry, harmonics, music therapy, the monochord, geometry, arithmetic, and cosmology. The school was destroyed by a rejected and disgruntled follower who led a popular revolt against the community around 500 B.C.E. Among the important followers of Pythagoras were Philolaus (b. 474 B.C.E.) and Archytas of Tarentum, an important influence on Plato.

The Pythagorean doctrine is based on these three principles:

- Ideas: matter is attracted to absolute forms, or ideas, which have an existence of their own. Mathematics is the study of these forms.
- Souls: an animal has an immortal soul, which reincarnates (transmigrates) after death, until a state of perfection is attained.
- Harmony: ideas and souls are related by sympathy, resonance, or musical ratio.

We may recognize the Pythagorean theory of reincarnation as derived from the Egyptian. The idea of the world soul evolved in this community.

Socrates (479–399 B.C.E.) was the agent of a major shift in which philosophy turned from nature (or physis) to human life. Also, he is considered among the first to emphasize the concept of the world soul.

Plato (429–347 B.C.E.) synthesized Pythagoras and Socrates. First he became a follower of Socrates. He had the genius to grasp Socrates' meaning, and to present it brilliantly in a series of ten dialogues. Around 390 B.C.E., Plato visited western Greece (Southern Italy and Sicily), encountered Pythagorean communities, met Archytas of Tarentum, the great Pythagorean, and adopted Pythagoreanism as a second influence. Platonism consists in the joining of these two streams, the Socratic and the Pythagorean. In 387 B.C.E., Plato created his school in Academe, a suburb of Athens.

Plato expanded the teaching of Socrates on the perfection of the soul into a complete system. In this system, morals and justice are based on absolute ideas. Wisdom consists of knowledge of these ideas, and philosophy is the search for wisdom. In fourteen more dialogues, Plato elaborated this unified system.

Plato's theory of soul is set out primarily in six of the dialogues: *Phaedo, Republic II,* and *Phaedra,* of the middle group of dialogues, 387–367

B.C.E.; *Timaeus,* around 365 B.C.E., which divides the middle and last groups; and *Philebus* and *Laws,* of the last group, 365–347 B.C.E. The development of the individual soul is given in the three middle dialogues.

The *Phaedo* is a long and detailed examination of the individual soul, its immortality, and reincarnation, given by Socrates on the day of his death sentence. The *Republic* describes Plato's mathematical curriculum for the Academy: arithmetic, plane geometry, solid geometry, astronomy, and music. At the end is the *Tale of Er,* which details the reincarnation process of the individual soul, as told by an eye witness. In the *Phaedrus,* Socrates and Phaedrus discourse on love, and on rhetoric. To understand divine madness, one must learn the nature of the soul. Soul is always in motion, and is self-moving, and therefore is deathless. Then begins the important metaphor of the chariot: two winged horses and a charioteer. This metaphor of the soul is used to explain divine madness, and the dynamics of reincarnation.

The world soul is developed in the later three dialogues. The *Timaeus* is a discussion of four persons: Socrates, Timaeus, Critias, and Hermocrates. It begins with a review by Socrates of a discussion on the preceding day. This concerned the constitution of the ideal State and its citizens. Then Critias tells the famous story of Atlantis, which was told to his great-grandfather by Solon, one of the seven sages. Then Timaeus is asked to begin the feast with a description of the creation of the universe. He tells how God, because he was good, made the world after an eternal pattern. He brought order into the world, and soul and intelligence. The world is composed of fire and earth. Being solids, these two elements require two more, water and air, to bind them. The world is a sphere with the soul in the center. The gods made man and the lower animals, and God made the human souls of the same four elements as the body of the universe, along with part of the soul of the universe. Then he set in motion the process of incarnation and reincarnation of these human souls in mortal bodies. The created gods make these mortal bodies of the four elements. As a person becomes a rational creature through education, his human soul moves in a circle in the head (a sphere) within his mortal body.

The *Philebus* is a lecture by Socrates on wisdom and pleasure. Along the way, he introduces the world soul as the source of individual souls. The *Laws* is the last of Plato's writings. It is a long dialogue of three older men, and is unique in that Socrates is absent. The actions of the world soul are discussed in detail.

3. The Stoics, 300 B.C.E. From Plato and Aristotle and their followers came the Stoics, the leading school for 500 years. Among other ideas, they further developed the logos concept of Heraclitus, Aristotle, and Philo. This became an element in the Neoplatonic cosmology of Plotinus. Logos has many meanings. Cognate of the verb legein, "to say," it may mean "language, speech, expression, explanation, formula, purpose, rational basis, plan."

Following Aristotle, the Stoics adopted two principles, or *archai:* one active, the other passive. According to David E. Hahm, in his *The Origins of Stoic Cosmology,* these are body and soul, or matter and logos. For the Stoics, logos makes the world by giving form to matter in a dynamical process. Like Plato, the Stoics believed that the cosmos was a living being, with a world soul.

4. Plotinus, 250 C.E. The main stimuli for the Neoplatonism of Plotinus (204–270) were Plato, the Middle Platonists, and, to a lesser extent, the Stoics. From Plato came Plotinus' main cosmology of the three primal hypostases: the *One,* the *Intelligence,* or *Intellectual Principle,* and the *World Soul.* For Plotinus the logos was a supplementary structure that intertwined the three hypostases. He defined it as "a power that acts upon matter, not conscious of it, but merely acting upon it."

This Neoplatonic cosmology, although further developed by Iamblichus, Proclus, and others, may be regarded as the main trunk of the Western Esoteric Tradition.

5. Spanda, 800 C.E. The Axial Age in India is represented by the older Upanishads, central to Vedanta, and the Yoga schools of ancient Sanskrit philosophy. In the middle ages, many schools diverged. One of these, Kashmiri Shaivism, developed a doctrine of vibration, given in the basic texts: *Siva Sutras* and *Spanda-Karikas.*

In his *Spanda-Karikas: The Divine Creative Pulsation,* Jaideva Singh notes that *spanda* means "some sort of movement." (From the Sanskrit dictionary: "throbbing, trembling, oscillation, vibration, or pulsation.") The text of the *Spanda-Karikas* suggests a sophisticated field concept applied to consciousness. This may have evolved from the older Akasha concept of the Upanishads.

6. *Al-Kindi, 850 c.e.* Al-Kindi (805–873) was an Islamic heir of Plato and the Neoplatonists. For him, the world soul was an emanation from the One, as light from the Sun. His astrological work, *De radiis,* was an important influence on the Western scientists Roger Grosseteste (1168–1253), Roger Bacon (1214–1294), Marsilio Ficino (1433–1499), and John Dee (1527–1609). *De radiis* presented an astrological theory based on rays from the planets. Everything radiates, and space is full of these radiations.

7. *Roger Bacon, 1267 c.e.* In *De multiplicatione specierum,* around 1267, Bacon presented a doctrine of the physics of light. He was inspired by Plotinus, Al-Kindi, and Roger Grosseteste. This doctrine survived for three centuries, and ended with John Dee. The word *species* meant the "likeness" of any object, transmitted through any media.

8. *Ficino, 1400 c.e.* Ficino's originality derived from a thoroughgoing syncretism of pagan and Christian elements, effected under the impulse of Plato, Plotinus, Proclus, the *Hermetica,* the Areopagite, Augustine, and Aquinas, to name only his primary wells of inspiration. Among the facets of this syncretism were:

- orphic music, music therapy (Ficino's personal practice)
- astrology (astrological psychology)
- magic, psychology

He was heir to the long line of astrological magic of Synesius, Proclus, Macrobius, and Al-Kindi, and was followed by Bruno and Agrippa. His cosmological model combined Neoplatonic and Christian elements, and set the foundation for the whole of Renaissance philosophy. It may be summarized in the following table.

Collective	Individual	Individual
	Discarnate	*Embodied*
The One *(ton het)*		
The Intelligence *(nous)*	Reason	Ideas
The Soul *(psyche)*	Angels	Individual soul (incl. mind)
Spirit *(pneuma)*	Stars	Individual spirit
Nature *(physis)*	Matter	Body

The One is the undivided source of everything. The Intelligence (Cosmic Mind) contains Plato's ideas, the archetypes and blueprints for creation. The Soul has three parts (rational, sensitive, and vegetative) and gives rise to individual minds, both human and angelic. Reason communicates between the Intelligence and the Soul, and Spirit (astral matter) intermediates between the World Soul and Nature, the created universe of matter, energy, and life.

Ficino's astrological magic, psychology, and medical practice were based on his understanding of Spirit, and its relation to the stars and planets. They have a contemporary revival in the work of James Hillman and Thomas Moore.

9. *Kepler, 1600 C.E.* Johannes Kepler (1571–1630) was a contemporary of: John Dee (1527–1609), the last emanationist; William Gilbert (1544–1603), the scholar of magnetism, the first field of modern physics; Giordano Bruno (1548–1600), the champion of the cosmos as an infinite plenum; and Galileo (1564–1642), the first modern dynamicist. In his work on elliptical orbits of the planets (especially Mars), Kepler proposed a theory of universal gravitation, the second field of modern physics. In his explanation of noncircular motion, he actually changed the word *spirit* (as in angelic influence) to *force* (that is, mechanism) in the manuscript for his most important work, *Astronomia Nova,* of 1609. And here we may locate the death of the world soul, concomitant with the birth of modern physics.

THE REBIRTH OF THE WORLD SOUL

The *individual soul* has been with us at least since 2500 B.C.E. But we have argued that the *world soul* emerged into documented literature with the Pythagoreans, around 500 B.C.E., and died with Kepler, around 1600 C.E., along with the birth of modern science. It has been missed. The support for our common sense of the coherence of all and everything has been lacking since modern science became our theology and cosmology. Calls for a renewed foundation for the cosmos are now multiplying, as the books of Rupert Sheldrake, Fritjof Capra, Dean Radin, and Ervin Laszlo, among others, testify.

10. Sheldrake, 1981. In his first book, *A New Science of Life: The Hypothesis of Formative Causation,* of 1981, Rupert Sheldrake begins with a consideration of unsolved problems of biology, in the areas of behavior, evolution, the origin of life, parapsychology, and so on. He delineates three levels of wholism: mechanism, vitalism, and organicism. We may relate these, respectively, to Nature, Spirit, and World Soul levels of the table above.

Building on the twentieth-century organismic ideas of Whitehead, Smuts, Waddington, and others, Sheldrake poses the existence of non-energetic fields, called morphogenetic fields, which direct the emergence of form in complex systems of all kinds. In the contexts of physics, chemistry, biology, and the social sciences, these may be called morphic fields, mental fields, family fields, and so on. Although non-energetic, these fields may have measurable effects on energetic systems. Sheldrake describes the effect of a morphogenetic field on an energetic system metaphorically as *morphic resonance.* His *hypothesis of formative causation* proposes that these fields evolve from unknown seeds called morphogenetic germs. Then they evolve their structures from previous similar systems; the past intervenes in the present; morphogenetic fields have memory.

In terms of the premodern cosmologies described above, we may locate Sheldrake's morphogenetic fields in the World Soul, while the morphogenetic germs reside in the Intelligence. The entire paradigm is organismic.

11. Capra, 1996. In his 1996 book, *The Web of Life: A New Scientific Understanding of Living Systems,* Fritjof Capra synthesizes the general systems theory of Ludwig von Bertalanffy, the cybernetics of Gregory Bateson and Humberto Maturana, and the environmental perspective of deep ecology into a top-down understanding of our biosphere. There is no appeal to organismic fields.

In the 2003 sequel, *The Hidden Connections: Integrating Biological, Cognitive, and Social Dimensions of Life into a Science of Sustainability,* Capra extends his biospheric big picture to include the social and economic spheres. His fully connected world-system is consistent with a vitalistic paradigm.

12. Laszlo, 2003–2005. In his 2003 book, *The Connectivity Hypothesis: Foundations of an Integral Science of Quantum, Cosmos, Life, and Consciousness,* Ervin Laszlo summarizes all the most important evidence for interconnecting fields, and also proposes a new physical field, the quantum vacuum field, as the carrier wave for the integrity of the cosmos. Unlike Sheldrake and Capra, Laszlo has insisted on a single, unique field as the basis of connectedness on all levels: physical, biological, social, and universal: truly an integral theory of everything.

In 2004, in his book *Science and the Akashic Field,* Laszlo proposes the quantum vacuum as the modern version of the Sanskrit *akasha,* an ether with memory. And in the present book, *Science and the Reenchantment of the Cosmos,* he develops the worldview and ethical implications of the Akashic or "A-field." This field, as Sheldrake's morphogenetic field, may be associated with the Intelligence of the Platonic cosmology as outlined above.

CONCLUSIONS

The New Sacred Math. All three of these pioneers of spiritual revival are dynamical-literate. They use the concepts of the new sacred math—catastrophe theory, chaos theory, bifurcations, neural networks, complex dynamical systems, emergence, complexity, agent based models, and the like—to formulate and evolve their ideas.

These new developments in computational mathematics provide the means for modeling and simulation of universal fields of information without the assumption of physical energy fields and specific communication means. For all these reasons we believe that the world soul is now being reborn from the ashes of inadequate hypotheses, such as the mechanical clockwork universe. A renaissance of premodern cosmology emerges in our postmodern culture, as the degenerate poverty of the received paradigm of modern science turns to compost.

We Are Not Alone

A Concluding Note

Jane Goodall

HOW WE BECAME THE "CONSCIOUS PART OF THE COSMOS"

I have spent forty-five years of my life learning about our closest living relatives, the chimpanzees. If, as Ervin Laszlo suggests, the human is "a *conscious* part of the cosmos, a being through which the world comes to know itself," the chimpanzee can be seen as a being through which we humans can come to better know ourselves, better understand how we acquired our state of consciousness, and better appreciate our relationship with the rest of the animal kingdom. And even, perhaps, help us to feel more at home in the universe, in leading-edge science's reenchanted cosmos.

Over the past four decades we have learned a great deal about chimpanzees. And what stands out is how like us they are, in so many ways. They have a complex repertoire of calls, postures, and gestures, which they use to communicate information about what is going on within and around them, and these include kissing, embracing, holding hands, swaggering, tickling. They use and make tools—an ability once thought to differentiate humans from the rest of the animal kingdom. Moreover, tool-using skills differ from one population to the next; as they are passed from one generation to the next through observation and imitation, these traditions can be labeled as primitive cultures.

It has been shown that chimpanzees are capable of self-recognition, abstraction, and generalization, and cross-modal transfer of information. They can learn over 400 of the signs of American Sign Language (ASL), use them to communicate with each other, invent signs for new objects, and clearly understand their meaning. They can accomplish many tasks using computer touch pads or cursors.

Chimpanzees show emotions similar to those we label happiness, sadness, fear, anger, and so on, and can experience mental as well as physical suffering. They also have a sense of humor. We have documented affectionate and supportive bonds between family members, which may persist through a life of sixty or more years. They care for each other and are capable of true altruism. Sadly, also like us, they have a dark side: they are aggressively territorial, and may perform acts of extreme brutality and even wage a kind of primitive war.

Human language. Clearly, then, there is no sharp line dividing humans from the rest of the animal kingdom. It is a very blurred line, and differences are of degree rather than of kind. Indeed, chimpanzees appear to stand at the very threshold of a human type of consciousness. What propelled our earliest ancestors across this threshold? What led to the development of an intellect that dwarfs that of even the most intelligent chimpanzee? Probably many factors were involved, but of crucial importance, I believe, is that at some point in our evolution we developed a sophisticated spoken language. Chimpanzees, as we have seen, have the cognitive ability to *learn* a human-type language. But, so far as we know, we are the only species in which individuals are able to share information about objects and events not present, learn from the distant past, and make plans for the far away future. Above all we can discuss ideas so that they can grow and change from the collective wisdom of a group. And today, thanks to the marvels of modern electronic communication, we can tap into the collective wisdom of people and cultures from around the world.

Religious behavior. From at least Neanderthal times, and probably before, humans everywhere have worshipped their gods. Religious, spiritual beliefs have sometimes endured through half a century or so of intense persecution—we only need to think of the suffering of the great Christian martyrs, the spiritual leaders and medicine men of the indigenous peoples who continued to practice their religious ceremonies secretly, the survival of Christianity through forty-five years of the communist regime in Eastern Europe. Without language, could we have developed a belief system? Is it possible that chimpanzees, without such a language, could show even the precursors of religious behavior? I think so.

Deep in the forest of Gombe is a spectacular waterfall. Sometimes as chimpanzees approach, and the roar of falling water gets louder, their pace quickens, and their hair bristles with excitement. When they reach the stream they perform magnificent displays, standing upright, swaying rhythmically from foot to foot, stamping in the shallow, rushing water, picking up and hurling great rocks. Sometimes they climb up the slender vines that hang down from high above and swing out into the spray of the falling water. This "waterfall dance" may last for ten or fifteen minutes and afterward a chimpanzee may sit on a rock, his eyes following the falling water. What is it, this water? It is always coming, always going—yet always there.

Do the chimpanzees feel an emotion similar to wonder or awe? And if they had a sophisticated spoken language, if they could *discuss* the feelings that trigger these magnificent displays, might that lead to the development of a primitive, animistic religion? The worship of inexplicable natural powers? For me the waterfall has always been the most spiritual place at Gombe, and we now know that it was considered a sacred place by the people who once lived there, a place where the medicine men performed ceremonies once a year. I wonder if they watched, entranced, the wild dancing of the chimpanzees?

Do chimpanzees have souls? Many people take comfort in thinking that apes, however like us they may be, at least do not have souls. But my years in the forest with the chimpanzees have led me to believe that if I have a soul, then so do they. Day after day I was alone in the wilderness, my companions the animals and the trees and the gurgling streams, the mountains and the awesome electrical storms and the star-studded night skies. I became one with a world in which, apart from the change from day to night, from wet season to dry, time was no longer important. Once or twice I attained a state of heightened awareness when it seemed that *self* was utterly absent: I and the chimpanzees, the earth and trees and air, seemed to merge, to become one with the life force of the forest. I became ever more attuned to the great Spiritual Power that I felt around me, as real as the beating of my heart.

And I came to believe that all living things posses a spark of that Spiritual Power. We humans, with our sophisticated, questioning minds and our spoken language, call that spark, in ourselves, a "soul." There is no real reason why it should not be the same for chimpanzees. But it is most unlikely that they, or any animal other than ourselves, care—or are capable of caring—as to whether or not they possess immortal souls!

MOVING TO HUMAN CONSCIOUSNESS

Why are we destroying Planet Earth? With or without a soul the chimpanzee, for me, seems to stand with one foot on the bridge over which, several million years ago, our ancestors crossed to become true humans, with spoken language and a heightened consciousness. "So what?" we may well ask. "Where has it got us?" Planet Earth has been overrun with these talking, conscious human primates. We have multiplied, often beyond the carrying capacity of the land we live on. As a result of poverty on the one hand, and the unsustainable life styles of the elite on the other, we are destroying the environment of our planet. We have used our brains to dominate nature, and to create awesome technology that, while it has often been used for great good, has also caused massive destruction. There is, surely, some justification for thinking that it might have been better if our ancestors had stayed with the chimpanzees on the other side of the bridge.

How has it happened? How is it possible that a conscious being, with a brain capable of the most extraordinary accomplishments, can be so shortsighted? What flaw is causing us to destroy our planet? To allow corporate greed to overcome our responsibilities to our children and all future generations? Why are we killing and maiming each other? How can we tolerate the unequal distribution of wealth, or the unsustainable use of finite natural resources that jeopardizes our very existence?

Admittedly we have inherited dark, selfish tendencies from our ancient primate heritage. But does that mean that violence and war, cruelty and greed, are inevitable? Not so—for our brains (unless we are mentally sick) are quite capable of overcoming lurking selfish genes. And remember, we have also inherited qualities of love, compassion, and altruism from our ancient primate heritage. There are those among us whose lives illuminate the limitless human potential for good.

But even if there is a sudden sea change, will it be in time? Will the conscious human be awakened before the destruction of life on Earth as we know it? I do believe that time is running out, but I still have hope.

WHY I BELIEVE THERE IS HOPE WE CAN COME HOME TO THE REENCHANTED COSMOS

My hope is based on four factors. Firstly, there is the extraordinary human brain. We are a problem-solving species. And now that we have finally admitted that we have, indeed, scarred planet Earth, so many brilliant and innovative technologies have been invented that will enable us to live in greater harmony with nature. Ordinary citizens are trying to leave lighter footprints as they move through life. And at least we know we must find solutions to population growth, poverty, and the unsustainable life styles of the elite.

Secondly, there is the resilience of nature. Habitats all but destroyed, species on the brink of extinction, people subjugated and all but broken, can so often recover—given time and a helping hand.

Thirdly, there is the energy, commitment, and enthusiasm of young people when they understand the problems and are empowered to act. The Jane Goodall Institute's program, Roots & Shoots, links young people from pre-school through university in over ninety countries, and empowers them to take action to help their human communities, animals, and the environment; and to learn more about those of other cultures, religions, and countries.

And fourthly, there is the indomitable human spirit. I know so many amazing people who tackle seemingly impossible tasks—and never give up. Who overcome seemingly insurmountable physical disabilities and lead lives that inspire us all. They have developed the compassionate, loving, and altruistic traits that both we and the chimpanzees have inherited from some ancient past. They illustrate our true human potential.

As we enter this twenty-first century, spirituality and science seem to be entering into a new relationship. As Ervin Laszlo shows us in this book, some of the latest thinking in physics, quantum mechanics, and cosmology are coming together in a new belief that Intelligence is involved in the formation of the universe, that there is Mind and Purpose underlying—and permeating—our existence. It is this that will, in the end, enable us to fully realize our human potential, to heal the harm that we have inflicted so that we may, once again, feel at home in a reenchanted cosmos.

SUMMING UP

Home Again in the Universe

As this book and the roundtable it includes has shown, findings at the cutting edge of contemporary science are reenchanting the cosmos. The newly reenchanted cosmos is pregnant with meaning for our life. It tells us that we are not strangers in the universe, but an integral part of it. It provides a new meeting ground between empirical science and intuitive spirituality.

Spirituality is based on timeless intuitions about the deeper or higher spheres of reality and it is essentially unchanged over the ages. Science, however, is—or should be—essentially an open enterprise. At its best it is not only a collection of abstract formulas, and not just a wellspring of technology; it is a source of insight into *what* there is in the world, and *how* things are in the world. By this token science is part of the perennial human quest for meaning and understanding. It is capable of change and renewal, and indeed it has changed fundamentally in the course of the twentieth century. In the first decade of the twenty-first century it is giving birth to an integral worldview. It is reenchanting the cosmos.

Unlike most people in contemporary society, leading scientists go beyond the narrow concept that sees spirituality as an indication of emotional and intellectual immaturity—due perhaps to lack of education, or to superstition and regression to magical and infantile thinking. More and more scientists seek answers to questions that were traditionally in the domain of spirituality. "What is the nature of the universe and what is our position in it? What does the universe mean? What is its purpose? Who are we and what is the meaning of our lives?" These questions were also asked by physicist David Peat. Science, he said, attempts to answer them since it has always been the province of the scientist to discover how the universe is constituted, how matter was first created, and how life began. Stephen Hawking shared this objective in his well-known books including *The Universe in a Nutshell.* So did physicist Sir Roger Penrose, who dedicated his newest book, *The Road to Reality: A Complete Guide to the Laws of the Universe,* to the quest of making fresh sense of the conceptual tangle of contemporary quantum physics. These scientists, while remaining dedicated to the rigorous method of science, are evolving a radically new view of the universe, and of the things that emerge and evolve in the universe.

There are, of course, scientists who work at the frontiers of their discipline but remain adamant that their findings are not relevant to questions of meaning. Some go as far as to say that the image of the universe disclosed in the theories of the natural sciences harbors no meaning for human life. The cosmological physicist Steven Weinberg maintains that the universe as a physical process is meaningless; the laws of physics offer no discernible purpose for humans. "I believe there is no point that can be discovered by the methods of science," he said in an interview. "I believe that what we have found so far—an impersonal universe, which is not particularly directed toward human beings—is what we are going to continue to find. And that when we find the ultimate laws of nature they will have a chilling, cold, impersonal quality about them." Meaning, according to Weinberg, resides in the human mind alone: the world itself is impersonal, without purpose or intention. Finding meaning in the universe is to make the error of projecting one's own mind and personality into it.

This view would make us strangers in the world. If mind, consciousness, and meaning were uniquely human phenomena, we would live in a universe devoid of the very qualities we ourselves possess, and our alienation would open the way to the blind exploitation of nature. If we arrogate all mind to ourselves, said ecologist Gregory Bateson, we see the world as mindless and therefore as not entitled to moral or ethical consideration. "If this is your estimate of your relation to nature and you have an advanced technology," he added, "your likelihood of survival will be that of a snowball in hell."

Is there truly no meaning to the universe? If so, we would be disoriented, deeply in trouble. Finding meaning in life and in nature is essential for human existence; the search for it is as old as civilization. For as long as people have looked at the sun, the moon, and the starry sky above, and at the seas, the rivers, the hills, and the forests below, they have wondered where it all came from, where it all is going, and what it all means. Traditional and non-Western peoples found answers, even if their answers were based on intuition rather than observation and experiment. They lived in a meaningful, ensouled universe and did not arrogate all mind to themselves. Finding themselves at home in the universe, they took care not to damage or destroy their home. Many such cultures endured for thousands of years.

The mechanistic worldview that took hold in Europe at the dawn of the Modern Age could not provide assurance that the world beyond the human mind has a meaning of its own. Following Newton, the universe was seen as a divinely designed clockwork set in motion by a prime mover and running harmoniously through all eternity. It operated according to rigorous laws. A knowledge of these laws would enable the human mind to know all things past, present, and future. In this universe the place of God was restricted to providing the original impetus that set the world in motion, and even this role was questioned. The French mathematician Laplace is reputed to have said to Napoleon that God is a hypothesis for which there is no longer any need.

As science joined with handicraft to produce a rapidly growing storehouse of mechanical innovations, nature was no longer considered of value in itself; it was primarily a resource for satisfying people's needs and wants. In the words of Francis Bacon, humans were free to wrest nature's secrets from her bosom for their own benefit. The progressive exploitation of the resources of nature, and therewith the increasing degradation of the life-support systems of the Earth's biosphere, was a logical consequence of this belief.

But we must not underestimate the ability of people, even of modern people, to find themselves at home again in nature and the universe. As Jane Goodall affirmed in the above roundtable on science and spirituality, there are good reasons to believe that we shall find our way home again. More and more people are seeking a meaningful concept of the world. The U.S. opinion survey *In Our Own Words* has put a quantitative figure to this search: it has found that 57 percent of Americans agree that a "global awakening to higher consciousness" is taking place. This hope and belief is evident in the flare-up of interest in the more serious kinds of science fiction, and in the great and at first surprising popularity of books, articles, and films that straddle the gap between science and spirituality—such as Fritjof Capra's *Tao of Physics,* Gary Zukav's *The Dancing Wu-Li Masters,* and more recently the film *What the Bleep Do We Know?* More and more people are attempting to move beyond the depressive futility of a disenchanted universe; more and more are ready to question whether we are truly the products of an accidental configuration of genes, resulting from chance mutations in our genome. In our day more and more would be ready to contest Bertrand Russell's famous declaration: "No fire, no heroism, no intensity of thought and feeling, can preserve an individual life beyond the grave . . . all the labors of the ages, all the devotion, all the inspiration, all the noonday brightness of human genius, are destined to extinction in the vast death of the solar system, and the whole temple of man's achievement must inevitably be buried beneath the debris of a universe in ruins."

Despite the outdated worldview most people associate with modern science, and notwithstanding the lingering skepticism of some scientists, science at its cutting edge rejects the idea of a disenchanted universe heading ineluctably toward a desolate destiny. The new cosmology discovers a world where the universe does not end in ruin, and the new physics, the new biology, and the new consciousness research recognize that we are part of an instantly and enduringly interconnected integral reality. Embracing the insights conveyed by these sciences, we and our fellow inhabitants of the planet need no longer feel ourselves strangers in the universe. We can come home again to a reenchanted cosmos.

Biographical Notes

ERVIN LASZLO has a Ph.D. from the Sorbonne and is the recipient of four honorary Ph.D.'s, the Japan Peace Prize (Goi Prize), and other distinctions. He was nominated for the Nobel Peace Prize in 2004 and renominated in 2005. Formerly Professor of Philosophy, Systems Science, and Futures Studies in various universities in the U.S., Europe, and the Far East, Laszlo is the author or coauthor of forty-five books translated into as many as twenty languages, and the editor of another twenty-nine volumes including a four-volume encyclopedia. He is Founder and President of The Club of Budapest, Founder and Director of the General Evolution Research Group, President of the Private University for Economics and Ethics, Fellow of the World Academy of Arts and Sciences, Member of the International Academy of Philosophy of Science, Senator of the International Medici Academy, and Editor of the international periodical *World Futures: The Journal of General Evolution.*

STANLEY KRIPPNER is professor of psychology at Saybrook Graduate School and Research Center, San Francisco, California, U.S.A. He is the recipient of the American Psychological Association's 2002 Award for Distinguished Contributions to the International Advancement of Psychology, and during the same year was given the Dr. J. B. Rhine Award for Lifetime Achievement in Parapsychology. Stanley Krippner is the co-editor of *Varieties of Anomalous Experience,* published by the American Psychological Association in 2000, and *The Psychological Impact of War Trauma on Civilians,* published by Greenwood in 2003. His coauthored books include *Dream Telepathy, Extraordinary*

Dreams, and *Becoming Psychic*. He is a former president of the International Association for the Study of Dreams, the Association for Humanistic Psychology, and the Parapsychological Association.

BRIAN A. CONTI is the founder and owner of BAC Gems Company in West Greenwich, Rhode Island, a consulting and appraisal service. He has published several articles on gemology in trade journals and is a long-time student of the Kabbalah and various other metaphysical systems.

ELISABET SAHTOURIS, evolution biologist, futurist, author, professor, and business consultant, taught at MIT and the University of Massachusetts, in the Bainbridge Graduate Institute's MBA program on sustainable business, and is a fellow of the World Business Academy. She has been a U.N. consultant on indigenous peoples and a NOVA/HORIZON TV writer. Her venues have included The World Bank, EPA, Boeing, Siemens, Hewlett-Packard, Tokyo Dome Stadium, Australian National Government, Sao Paulo's leading business schools, State of the World Forums (New York and San Francisco), and World Parliament of Religion, South Africa. Her books include *EarthDance: Living Systems in Evolution, A Walk Through Time: From Stardust to Us,* and *Biology Revisioned* (with Willis Harman).

CHRISTIAN DE QUINCEY is Professor of Philosophy and Consciousness Studies at John F. Kennedy University, and founder of The Visionary Edge, a media network and production company committed to transforming global consciousness by transforming mass media. He is author of the award-winning book *Radical Nature: Rediscovering the Soul of Matter* (2002), as well as *Radical Knowing: Understanding Consciousness through Relationship* (2005), and the novel *Deep Spirit: Lessons from the Noetic Code* (spring, 2006).

EDGAR MITCHELL, Captain, U.S. Navy (retired); has a Sc.D. from the Massachusetts Institute of Technology in 1964. His honorary doctorates include Carnegie Mellon University, 1971; New Mexico State University, 1971; University of Akron, 1979; and Embry-Riddle University, 1996. Mitchell was captain of the Apollo 14 mission and in February 1971 was the sixth man to walk on the moon. He founded the Institute of Noetic Sciences in 1973 and co-founded the Association of Space Explorers in 1984. He is the author of *Psychic Explo-*

ration: A Challenge for Science (1974), and *The Way of the Explorer,* with Dwight Williams (1996). He lectures worldwide and consults with major public organizations and business companies.

STANISLAV GROF is a psychiatrist with more than fifty years of experience in research on non-ordinary states of consciousness. His past appointments include Principal Investigator in a psychedelic research program at the Psychiatric Research Institute in Prague, Czechoslovakia, Chief of Psychiatric Research at the Maryland Psychiatric Research Center, Assistant Professor of Psychiatry at the Johns Hopkins University in Baltimore, Maryland, and Scholar-in-Residence at the Esalen Institute in Big Sur, California. Currently he is Professor of Psychology at the California Institute of Integral Studies (CIIS) and Pacifica Graduate Institute in Santa Barbara, California. Stanislav Grof conducts professional training programs in holotropic breathwork and transpersonal psychology, and gives lectures and seminars worldwide. He is one of the founders and chief theoreticians of transpersonal psychology and the founding president of the International Transpersonal Association (ITA). His publications include over 100 papers in professional journals and the books *Beyond the Brain, The Adventure of Self-Discovery, LSD Psychotherapy, Books of the Dead, The Cosmic Game, The Consciousness Revolution* (with Ervin Laszlo and Peter Russell), *Psychology of the Future, Beyond Death,* and *The Stormy Search for the Self* (the last two with Christina Grof).

PETER RUSSELL has degrees in theoretical physics, experimental psychology, and computer science from the University of Cambridge, England. In India, he studied meditation and Eastern philosophy, and on his return took up research into the psychophysiology of meditation at the University of Bristol. He was one of the first to introduce human potential seminars into the corporate field, and for twenty years worked with major corporations on creativity, learning methods, stress management, and personal development. His books include *The Global Brain Awakens, Waking Up in Time,* and most recently *From Science to God.*

H. JOHANNES WITTEVEEN, PH.D., pursued a successful career as an economist, during which he held the posts of Professor of Economics in Rotterdam, Minister of Finance of The Netherlands and Managing Director of the

International Monetary Fund. He has made a life-long study of universal Sufism, and has continued to give time to development of the inner life alongside his economic activities. He is Vice President of the International Sufi Movement.

SWAMI KRIYANANDA (J. Donald Walters) has been a direct disciple since 1948 of Paramhansa Yogananda, author of *Autobiography of a Yogi.* He is himself author of some eighty-five books published in twenty-seven languages. Kriyananda has also created a one-year correspondence course, "Material Success Through Yoga Principles," and 365 twenty-minute segments of a television program called "The Way of Awakening." The program appears daily throughout the year on two television stations in India and is available in more than one hundred other countries. In addition, Kriyananda has composed over four hundred pieces of music, both vocal and instrumental. He has received several awards, including the 2005 international "Goodness" Award in Milan, Italy. Swami Kriyananda is the founder of seven spiritual communities in America and in Italy, with an eighth being established in India.

IRENE VAN LIPPE-BIESTERFELD, Princess of the Netherlands, works in the field of personal empowerment, as a social educator in Europe, South Africa, and the United States. She worked in the political field in Spain; she was active in feminism and the antinuclear movement; in cultural diversity in the Netherlands (co-founder of the Isis Transcultural Leadership, Business in a Changing World, 1983); and in the natural environment (Founder/President of the Foundation Lippe-Biesterfeld Nature-College, 2001). She has written several books. *Dialogue with Nature* (1995) was a bestseller in the Netherlands and was translated into five languages. Her latest book is *Science, Soul, and the Spirit of Nature.* Irene van Lippe-Biesterfeld is the mother of four children.

EWERT COUSINS is the General Editor of the twenty-five-volume series *World Spirituality: An Encyclopedic History of the Religious Quest* and Editorial Consultant of the series *Classics of Western Spirituality.* He is a translator of the texts of Bonaventure and an interpreter of his thought, former president of the American Teilhard Association, and the author of several books, including *Global Spirituality: Toward the Meeting of Mystical Paths.* He is Professor Emeritus of Theology, Fordham University, and was Visiting Professor at Columbia University, New York University, and Hebrew University, Jerusalem.

RALPH H. ABRAHAM has been Professor of Mathematics at the University of California at Santa Cruz since 1968. He received a Ph.D. in Mathematics at the University of Michigan in 1960, and taught at Berkeley, Columbia, and Princeton before moving to Santa Cruz. He has held visiting positions in Amsterdam, Paris, Warwick, Barcelona, Basel, and Florence, and is the author of more than twenty texts, including eight books currently in print. In 1975 he founded the Visual Mathematics Project at the University of California at Santa Cruz, which became the Visual Math Institute in 1990, with its popular Web site since early 1994. He has performed works of visual and aural mathematics and music (with Ami Radunskaya and Peter Broadwell) since 1992.

JANE GOODALL is known for her landmark study of chimpanzees in Gombe National Park, Tanzania. In 1977, she established the Jane Goodall Institute (JGI), which supports the Gombe research and is a global leader in the effort to protect chimpanzees and their habitats. With nineteen offices around the world, the Institute is widely recognized for innovative, community-centered conservation and development programs in Africa and a global youth program, Roots & Shoots, which currently operates in eighty-seven countries. She devotes virtually all of her time to advocating on behalf of chimpanzees and the environment, traveling nearly three hundred days a year. Her many honors include the Medal of Tanzania, Japan's Kyoto Prize, the Benjamin Franklin Medal in Life Science, and the Gandhi/King Award for Nonviolence. In April 2002, Secretary-General Kofi Annan named Jane Goodall a United Nations "Messenger of Peace."

References and Further Reading

Chapter One

Mae-Wan Ho's recent findings on the coherence of the living organism are succinctly stated in *The Rainbow and the Worm: The Physics of Organisms* (Singapore and London: World Scientific, 1993).

Works of general interest regarding the new cosmology include:

John Gribbin, *In the Beginning: The Birth of the Living Universe* (New York: Little, Brown & Co., 1993); Alan H. Guth, *The Inflationary Universe: The Quest for a New Theory of Cosmic Origins* (New York: Perseus Books, 1997); Craig J. Hogan, *The Little Book of the Big Bang* (New York: Springer Verlag, 1998); Menas Kafatos and Robert Nadeau, *The Conscious Universe: Part and Whole in Modern Physical Theory* (New York: Springer Verlag, 1990, 1999); Eugene F. Mallove, "The self-reproducing universe," *Sky & Telescope* 76:3 (September 1988); P. Peebles and E. James, *Principles of Physical Cosmology* (Princeton, N.J.: Princeton University Press, 1993); Martin Rees, *Before the Beginning: Our Universe and Others* (New York: Addison-Wesley, 1997); as well as John Leslie, *Universes* (London and New York: Routledge, 1989) and the multi-authored volume edited by Leslie, *Physical Cosmology and Philosophy* (New York: Macmillan, 1990).

The famous paper giving the surprising results of the EPR experiment is Alain Aspect and P. Grangier, "Experiments on Einstein-Podolski-Rosen-type correlations with pairs of visible photons" in *Quantum Concepts in Space and Time*, edited by Roger Penrose and C. J. Isham (Oxford: Clarendon Press, 1986).

The original reports on the landmark teleportation experiments are by M. D. Barret et al., "Deterministic quantum teleportation of atomic qubits," *Nature*

429 (June 17, 2004); and M. Riebe et al., "Deterministic quantum teleportation with atoms," *Nature* 429 (June 17, 2004).

Michael Behe gives his theory of the irreducible complexity of living systems in *Darwin's Black Box: The Biochemical Challenge to Evolution* (New York: Touchstone Books, 1998).

Experiments on the sensitivity of the genome to external influences are reported by Michael L. Lieber in "Environmentally responsive mutator systems: toward a unifying perspective," *Rivista di Biologia/Biology Forum* 91, no. 3 (1998); "Hypermutation as a means to globally restabilize the genome following environmental stress," *Mutation Research, Fundamental and Molecular Mechanisms of Mutagenesis* 421, no. 2 (1998); and in "Force and genomic change," *Frontier Perspectives* 10:1 (2001).

Details on the experiments on brain-wave synchronization among different individuals are reported by Nitamo Montecucco in *Cyber: La Visione Olistica* (Rome: Mediterranee, 2000) (in Italian).

Works describing the telepathic abilities of the Australian aborigines include: A. P. Elkin, *The Australian Aborigines* (Sydney: Angus & Robertson, 1942); Robert Lawlor, *Voices of the First Day: Awakening in the Aboriginal Dreamtime* (Rochester, Vt.: Inner Traditions, 1991); and Marlo Morgan, *Mutant Message Down Under* (New York: HarperCollins, 1991).

Seminal works on telepathic ability in ordinary people are inter alia by M. A. Persinger and Stanley Krippner, "Dream ESP experiments and geomagnetic activity," *Journal of the American Society for Psychical Research* 83 (1989); and Harold Puthoff and Russell Targ, "A perceptual channel for information transfer over kilometer distances: historical perspective and recent research," *Proceedings of the IEEE* 64 (1976). Telepathy in twins is demonstrated in Guy Playfair, *Twin Telepathy: The Psychic Connection* (London: Vega Books, 2002).

The results of the U.S. opinion survey on common core values and beliefs of Americans was published by the Fund for Global Awakening: *What Brings Us Together: Presentation of The IOOW 2000 Research Program* (Point Reyes Station, Calif., 2001).

For an overview of Stanislav Grof's relevant findings, see his Comment in part three. Grof's principal published works include: *The Adventure of Self-Discovery* (Albany: State University of New York Press, 1988); *Psychology of the Future* (Albany: State University of New York Press, 2000); and coauthored with Hal Zina Bennett, *The Holotropic Mind* (San Francisco: Harper, 1993).

Chapter Two

Swami Vivekananda's celebrated opus is *Raja Yoga* (Calcutta: Advaita Ashrama, 1982). The citation is from chapter 3.

Radja Deekshithar's paper was published (in translation) in the Dutch magazine *BRES* 194 (1996).

The original report on inertia as a vacuum force is Bernhard Haisch, Alfonso Rueda, and Harold E. Puthoff, "Inertia as a zero-point-field Lorentz force," *Physical Review A*, 49:2 (1994). Further reports on vacuum interaction effects include the following papers by Harold Puthoff, "Ground state of hydrogen as a zero-point-fluctuation-determined state." *Physical Review D* 35, no. 10 (1987); "Source of vacuum electromagnetic zero-point energy," *Physical Review A* 40, no. 9 (1989); and "Gravity as a zero-point-fluctuation force," *Physical Review A* 39, no. 5 (1989, 1993).

The new discovery regarding gravity is reported in Dezsõ Sarkadi and László Bodonyi, "Gravity between commensurable masses," Private Research Centre of Fundamental Physics, *Magyar Energetika* 7, no. 2 (1999).

Astronaut Edgar Mitchell recounts his experiences and gives his conclusions in *The Way of the Explorer: An Apollo Astronaut's Journey through the Material and Mystical Worlds* (New York: Putnam, 1996). His pertinent views are summarized by Doris Lora in "A Dyadic Model of Reality," *Shift at the Frontiers of Consciousness, Journal of the Institute of Noetic Sciences* (December 2003–February 2004).

The Russian physicists' torsion wave theory is stated in A. E. Akimov and G. I. Shipov, "Torsion fields and their experimental manifestations," *Journal of New Energy* 2, no. 2 (1997); A. E. Akimov and V. Y. Tarasenko, "Models of polarized states of the physical vacuum and torsion fields," *Soviet Physics Journal* 35, no. 3 (1992); and G. I. Shipov, *A Theory of the Physical Vacuum: A New Paradigm* (Moscow: International Institute for Theoretical and Applied Physics RANS, 1998).

László Gazdag's writings include *Beyond the Theory of Relativity* (Budapest: Robottechnika, 1998); "Superfluid mediums, vacuum spaces," *Speculations in Science and Technology* 12, no. 1 (1989); and "Combining of the gravitational and electromagnetic fields," *Speculations in Science and Technology* 16, no. 1 (1993).

Chapter Three

Details of the amazing fine-tuning of the universal constants is described in John D. Barrow and Frank J. Tipler, *The Anthropic Cosmological Principle* (London and New York: Oxford University Press, 1986).

The quasi-steady-state cosmology is advanced in the technical paper by Fred Hoyle, G. Burbidge, and J. V. Narlikar, "A quasi-steady state cosmology model with creation of matter," *Astrophysical Journal* 410 (June 20, 1993).

The cyclic cosmology from Big Bang to Big Crunch and Big Bang again is put forward by Paul J. Steinhardt and Neil Turok, "A cyclic model of the universe," in *Science* 296 (2002).

Harlow Shapley gave his estimate of life-bearing plants in the book *Of Stars and Men* (Boston: Beacon, 1958); and Su-Shu Huang's estimate is reported in *American Scientist* 47 (1959).

Fred Hoyle gives the Rubik's cube example in *The Intelligent Universe* (London: Michael Joseph, 1983).

David Chalmers posed his "hard problem" in *The Conscious Mind: In Search of a Theory of Conscious Experience* (New York: Oxford University Press, 1996), and in "The puzzle of conscious experience," *Scientific American* 273, no. 6 (1995).

Jerry Fodor's comment is in his article "The big idea," *The New York Times Literary Supplement* (July 3, 1992).

Freeman Dyson's views on matter are in his *Infinite in All Directions* (New York: Harper & Row, 1988).

For Grof's findings, see his Comment in part three, as well as the references to his writings below and in chapter 2.

Simon Berkovitsch's estimate of the brain's information-storing capacity is in "On the information processing capabilities of the brain: shifting the paradigm," *Nanobiolog* 2 (1993); and Herms Romijn's estimate is in the paper "About the origins of consciousness: a new multidisciplinary perspective on the relationship between brain and mind," *Proc. Kon. Ned. Akad. v Vetensch* 1–2 (1977).

Chapter Four

Ian Stevenson reported on children's past-life stories in his books *Children Who Remember Previous Lives* (Charlottesville: University of Virginia Press, 1987); *Cases of the Reincarnation Type*, 4 vols. (Charlottesville: University of Virginia Press, 1975–83); and *Reincarnation and Biology: A Contribution to the Etiology of Birthmarks and Birth Defects,* 2 vols. (Westport, Ct.: Praeger, 1997). The story of Parmod Sharma comes from his *Twenty Cases Suggestive of Reincarnation* (Charlottesville: University of Virginia Press, 1966).

Cardiologist Pim van Lommel's report on consciousness in clinically brain-dead patients is on the Internet under the title "Consciousness and the brain: A new concept about the continuity of our consciousness based on recent scientific research on near-death experience," pimvanlommel@wanadoo.nl, 2004.

Sabine Wagenseil's report on her after-death communication experience is published in German under the title "Tod ist nicht tödlich: durchgaben über den Tod von einem Toten" (Death is not deadly: transmissions about death from a dead), in *Grenzgebiete der Wissenschaft* 51, no. 2 (2002).

The Transcommunication Study Circle of Luxembourg (CETL: Centre d'Etudes sur la Transcommunication—Luxembourg) published an overview of their "instrumental transcommunications" in *Conversations Beyond the Light,* by Pat Kubis and Mark Macy (Boulder, Colo.: Griffin Publishing, 1995).

For the after-death communication experiments of Dr. Botkin and colleagues see Allan Botkin and Craig Hogan, *Induced After-Death Communication: A New Therapy for Healing Grief and Trauma* (Charlottesville, Va.: Hampton Roads, 2005).

Chapter Five

William Clifford's statements are cited by Milo Wolf and Geoff Haselhurst in "Einstein's Last Question," *VIA: Journal of Integral Thinking for Visionary Action* 3, no. 1 (2005).

Albert Einstein gave his reconsidered views on space in "The Concept of Space," *Nature* 125 (1930); and Erwin Schrödinger's views on this topic are in *Schrödinger: Life and Thought* (London: Cambridge University Press, 1989).

Shahriar Afshar reported the revolutionary results of his split-beam experiments in the June 2004 issue of the British journal *New Scientists.*

For Milo Wolf's wave theory of matter, see inter alia his "Gravitation and Cosmology," in *From the Hubble Radius to the Planck Scale,* edited by R. L. Amoroso et al. (Amsterdam: Kluwer Academic Publishers, 2002).

J. Scott Russell's account of his encounter with a soliton was published by the British Association for the Advancement of Science in 1845 under the title "Report on Waves."

Home Again in the Universe

David Peat posed his questions about science and meaning in *Synchronicity: The Bridge Between Matter and Mind* (New York: Bantam Books, 1987); and Steven Weinberg made his skeptical comments in an interview: "Lonely planet," *Science and Spirit* 10, no. 1 (April–May 1999).

The quotation from Gregory Bateson is from his *Steps to an Ecology of Mind* (New York: Ballantine, 1972).

Bertrand Russell's oft-quoted remarks are from his paper, "A freeman's worship," in *The Basic Writings of Bertrand Russell 1903–1959,* edited by R. E. Egner and L. D. Dennon (New York: Simon & Schuster, 1960).

Roundtable References

Krippner and Conti References

Cardeña, E., S. J. Lynn, and S. Krippner. *Varieties of Anomalous Experience: Examining the Scientific Evidence.* Washington, D.C.: American Psychological Association, 2000.

Cheney, R. *Akashic Records: Past Lives and New Directions.* Upland, Calif.: Astara, 1996.

Laszlo, E. *The Interconnected Universe: Conceptual Foundations of Transdisciplinary Unified Theory.* London: World Scientific, 1995.

Laszlo, E. *Science and the Akashic Field: An Integral Theory of Everything.* Rochester, Vt.: Inner Traditions, 2004.

McTaggart, L. *The Field: The Quest for the Secret Force of the Universe.* New York: Quill/HarperCollins, 2002.

von Bertalanffy, L. *General System Theory: Essays on Its Foundation and Development.* Revised ed. New York: George Brazillier, 1968.

Sahtouris References

Sahtouris, Elisabet. *EarthDance: Living Systems in Evolution.* iUniverse.com, New York: Ingram, 2000 (distribution).

Maturana, Humberto R., and Francisco Varela. *The Tree of Knowledge: The Biological Roots of Human Understanding.* Boston: Shambhala, 1987.

de Quincey References

de Quincey, Christian. *Radical Nature: Rediscovering the Soul of Matter.* Montpelier, Vt.: Invisible Cities Press, 2002.

Laszlo, Ervin. *The Connectivity Hypothesis: An Integral Theory of Quantum, Cosmos, Life, and Consciousness*. Albany: State University of New York (SUNY) Press, 2003.

Mitchell References

Anandan, J. "The Geometric Phase." *Nature* 360, no. 26 (1992).

Bouwmeester, D., et al. *Nature* 390, no. 11 (December 1997).

Byrd, R. C. "Positive therapeutic effect of intercessory prayer in a coronary care population." *Southern Medical Journal* 81, no. 7 (1988).

Dossey, L. *Healing Words*. San Francisco: Harper San Francisco, 1993.

Dunne, B. J., and R. D. Nelson. *Journal of Scientific Exploration* 2 (1988):155–80.

Eberhard, P. H. "Bell's Theorem without Hidden Variables." *Il Nuovo Cimento* 38, B 1 (March 1977).

Frieden, B. Roy. *Physics from Fisher Information*. Cambridge: Cambridge University Press, 1998.

Haisch, B., A. Rueda, H. E. Puthoff. "Physics of the Zero-Point Field: Implications for Inertia, Gravity, and Mass." *Speculations in Science and Technology* 20 (1997).

———. "Advances in the Proposed Zero-Point Field Theory of Inertia." 34th AIAA Joint Propulsion Conference, AIAA paper no. 98 3143 (1998).

Hammeroff, S. R. "Quantum coherence in microtubules: A neural basis for emergent consciousness?" *Journal of Consciousness Studies* 1 (1994): 91–118.

Marcer, P. J. "Getting Quantum Theory off the Rocks: Nature, as we consciously perceive it, is quantum reality!" *Proc. 14 International Congress of Cybernetics,* Namur (Aug 21–25, 1995).

Marcer, P. J., and W. Schempp. "A mathematically specified template for DNA and the genetic code in terms of the physically realizable processes of quantum holography." *Proc. Greenwich Symposium on Living Computers*, edited by A. M. Fedorec and P. J. Marcer, 1996.

———. "Model of the Neuron Working by Quantum Holography." *Informatica* 21 (1997).

———. "The model of the prokaryote cell as an anticipatory system working by quantum holography." *International Journal of Computing Anticipatory Systems* (1998), CHAOS Vol. 2.

Mitchell, E. D., and D. Williams. *The Way of the Explorer*. New York: Putnam, 1996.

Penrose, R. *The Emperor's New Mind*. Oxford: Oxford University Press, 1999.

Pribram, K. H. "Quantum holography: Is it relevant to brain function?" *Information Sciences* 115 (1999).

Puthoff, H. E. "CIA-initiated remote viewing program at Stanford Research Institute." *Journal of Scientific Exploration* 10 (1996): 63–76.

Radin, Dean. *The Conscious Universe.* San Francisco: Harper, 1997.

Schempp, W. "Harmonic analysis on the Heisenberg group with application in signal theory." *Pitman Research Notes in Mathematics*, Series 14. London: Longman Scientific and Technical, 1986.

———. "Quantum Holography and Neurocomputer Architectures." *Journal of Mathematical Imaging and Vision* 2 (1992): 109–164.

Tiller, W. A. *Science and Human Transformation.* Calif.: Pavoir Publishing, 1997.

Utts, J. M. "Replication and meta-analysis in parapsychology." *Statistical Science* 6 (1991).

Van Flandern, T. "What the Global Positioning System tells us about Relativity." In *Open Questions in Relativistic Physics,* edited by F. Selleri. Montreal: Apeiton, 1998.

Grof References

Bohm, D. *Wholeness and the Implicate Order.* London: Routledge & Kegan Paul, 1980.

Campbell, J. *The Hero with A Thousand Faces.* Princeton, N.J.: Princeton University Press, 1968.

Corbin, H. "Mundus Imaginalis, Or the Imaginary and the Imaginal." *Working With Images*, edited by B. Sells. Woodstock, Ct.: Spring Publications, 2000, 71–89.

Capra, F. *The Tao of Physics.* Berkeley: Shambhala Publications, 1975.

Ferenczi, S. "Thalassa." *The Psychoanalytic Quarterly.* New York, 1938.

Grof, S. *LSD Psychotherapy.* Pomona, Ca.: Hunter House, 1980.

———. *Beyond the Brain: Birth, Death, and Transcendence in Psychotherapy.* Albany: State University of New York (SUNY) Press, 1985.

———. *The Adventure of Self-Discovery.* Albany: State University of New York (SUNY) Press, 1987.

———. *Psychology of the Future: Lessons from Modern Consciousness Research.* Albany: State University of New York (SUNY) Press, 2000.

Grof, S., and C. Grof. *Spiritual Emergency: When Personal Transformation Becomes a Crisis.* Los Angeles: J. P. Tarcher, 1989.

———. *The Stormy Search for the Self: A Guide to Personal Growth Through Transformational Crises.* Los Angeles: J. P. Tarcher, 1991.

Jung, C. G. *Symbols of Transformation.* Collected Works, vol. 5. Bollingen Series XX. Princeton N.J.: Princeton University Press, 1956.

———. *The Archetypes and the Collective Unconscious*. Collected Works, vol. 9, 1. Bollingen Series XX. Princeton, N.J.: Princeton University Press, 1959.

Laszlo, E. *The Creative Cosmos*. Edinburgh: Floris Books, 1993.

———. *The Interconnected Universe. Conceptual Foundations of Transdisciplinary Unified Theory*. Singapore: World Scientific Publishing, 1994.

———. *The Whispering Pond: A Personal Guide to the Emerging Vision of Science*. Boston: Element Books, 1999.

———. *The Connectivity Hypothesis: Foundations of an Integral Science of Quantum, Cosmos, Life, and Consciousness*. Albany: State University of New York (SUNY) Press, 2003.

———. *Science and the Akashic Field: An Integral Theory of Everything*. Rochester, Vt.: Inner Traditions, 2004.

Fodor, Nandor. *The Search for the Beloved: A Clinical Investigation of the Trauma of Birth and Prenatal Condition*. New Hyde Park, N.Y.: University Books, 1949.

Peerbolte, L. "Prenatal Dynamics." In *Psychic Energy*. Amsterdam: Servire Publications, 1975.

Pribram, K. *Languages of the Brain*. Englewood Cliffs, N.J.: Prentice Hall, 1971.

Prigogine, Ilya. *From Being to Becoming: Time and Complexity in the Physical Sciences*. San Francisco: W. H. Freeman, 1980.

Prigogine, Ilya, and Isabelle Stengers. *Order out of Chaos: Man's Dialogue with Nature*. New York: Bantam Books, 1984.

Rank, O. *The Trauma of Birth*. New York: Harcourt Brace, 1929.

Reich, W. *Character Analysis*. New York: Noonday Press, 1949.

———. *The Function of the Orgasm: Sex-Economic Problems of Biological Energy.* New York: Farrar, Strauss & Giroux, 1961.

Sheldrake, R. *A New Science of Life: The Hypothesis of Formative Causation*. Los Angeles: J. P. Tarcher, 1981.

Talbot, M. *The Holographic Universe*. New York: Harper Collins Publishers, 1991.

Wilber, K. *A Theory of Everything: An Integral Vision for Business, Politics, Science and Spirituality*. Berkeley: Shambhala Publications, 2000.

Russell References

Laszlo, Ervin. *The Connectivity Hypothesis: Foundations of an Integral Science of Quantum, Cosmos, Life, and Consciousness*. Albany: State University of New York (SUNY) Press, 2003.

Russell, Peter. *From Science to God: A Physicist's Journey into the Mystery of Consciousness*. Pub Group West, 2005.

Witteveen References

Khan, Hazrat Inayat. *The Mysticism of Sound and Music* Boston and London: Shambhala, 1996.

———. *The Sufi Message*, Fourteen Volumes. London and The Hague: East West Publications and New Lebanon, N.Y.: Omega Publications, 1990–1996.

Lippe-Biesterfeld References

Berry, Thomas. *The Great Work*. New York: Bell Tower, 1999.

Cousins References

Houston, Jean. *The Search for the Beloved: Journeys in Mythology & Sacred Psychology*. Los Angeles: J. P. Tarcher, 1987; New York: St. Martin's Press, 1997; *A Passion for the Possible*. San Francisco: Harper, 1997.

Houston, Jean, and Robert Masters. *Mind Games: The Guide to Inner Space*. Los Angeles: J. P. Tarcher, 1987; *Varieties of Psychedelic Experience*. New York: Holt, Rinehart and Winston, 1969; Rochester, Vt.: Park Street Press, 2000.

Teilhard de Chardin, Pierre. "Discovering the Divine in the Depths of Blazing Matter." In *Pierre Teilhard de Chardin: Writings*. Selected with an Introduction by Ursala King. Maryknoll, N.Y.: Orbis Books, 1999.

Abraham References

Allen, Don Cameron. *The Star-crossed Renaissance: The Quarrel about Astrology and Its Influence in England*. New York: Octagon Books, 1941/1966.

Allen, Michael J. B. *Marsilio Ficino and the Phaedran Charioteer*. Berkeley: University of California Press, 1981.

Capra, Fritjof. *The Web of Life: A New Scientific Understanding of Living Systems*. New York: Anchor, 1996.

———. *The Hidden Connections: Integrating Biological, Cognitive, and Social Dimensions of Life into a Science of Sustainability*. New York: Doubleday, 2002.

Couliano, Ioan P. *Eros and Magic in the Renaissance*. Chicago: University of Chicago Press, 1984/1987.

Damiani, Anthony J. *Astronoesis*. Burdett, N.Y.: Larson, 2000.

Deck, John. *Nature, Contemplation, and the One, a Study in the Philosophy of Plotinus*. Burdett, N.Y.: Larson, 1967/1991.

Dyczkowski, Mark S. G. *The Stanzas on Vibration: The Spandakarika with Four Commentaries*. Albany: State University of New York (SUNY) Press, 1992.

Gilbert, William. *On the Magnet*. New York: Basic Books, 1958.

Gregory, John. *The Neoplatonists, a Reader*. London: Routledge, 1991/1999.

Hahm, David E. *The Origins of Stoic Cosmology*. Columbus: Ohio State University Press, 1977.

Hamilton, Edith, and Huntington Cairns. *Plato, Collected Dialogues*. Princeton, N.J.: Princeton University Press, 1961.

Hillman, James. *Archetypal Psychology: A Brief Account, together with a Complete Checklist of Works*. Dallas: Spring Publications, 1983.

Kraut, Richard, ed. *The Cambridge Companion to Plato*. Cambridge: Cambridge University Press, 1992.

Laszlo, Ervin. *The Connectivity Hypothesis: Foundations of an Integral Science of Quantum, Cosmos, Life, and Consciousness*. Albany: State University of New York (SUNY) Press, 2003.

———. *Science and the Akashic Field: An Integral Theory of Everything*. Rochester, Vt.: Inner Traditions, 2004.

Lindberg, David C. *Theories of Vision from al-Kindi to Kepler*. Chicago: University of Chicago Press, 1976.

———. *Roger Bacon's Philosophy of Nature*. Oxford: Oxford University Press, 1983.

———. *Roger Bacon and the Origins of Perspectiva in the Middle Ages*. Oxford: Oxford University Press, 1996.

Moore, Thomas. *The Astrological Psychology of Marsilio Ficino*. Lewisburg, Penn.: Bucknell University Press, 1982.

Sandbach, F. H. *The Stoics*. New York: Norton, 1975.

Sheldrake, Rupert. *A New Science of Life: The Hypothesis of Formative Causation*. Los Angeles: J. P. Tarcher, 1981.

Singh, Jaideva. *Spanda-Karikas: The Divine Creative Pulsation*. Delhi: Motilal Banarsidass, 1980.

Turnbull, Grace H. *The Essence of Plotinus*. New York: Oxford University Press, 1948.

West, John Anthony. *The Traveler's Key to Ancient Egypt, A Guide to the Sacred Places of Ancient Egypt*, 2nd edition. Wheaton, Ill.: Quest Books, 1985/1995.

Index

BOOKS OF RELATED INTEREST

Science and the Akashic Field
An Integral Theory of Everything
by Ervin Laszlo

Science, Soul, and the Spirit of Nature
Leading Thinkers on the Restoration of Man and Creation
by Irene van Lippe-Biesterfeld with Jessica van Tijn

The Rebirth of Nature
The Greening of Science and God
by Rupert Sheldrake

A New Science of Life
The Hypothesis of Morphic Resonance
by Rupert Sheldrake

Matrix of Creation
Sacred Geometry in the Realm of the Planets
by Richard Heath

Genesis of the Cosmos
The Ancient Science of Continuous Creation
by Paul A. LaViolette, Ph.D.

Alternative Science
Challenging the Myths of the Scientific Establishment
by Richard Milton

The Biology of Transcendence
A Blueprint of the Human Spirit
by Joseph Chilton Pearce

Inner Traditions • Bear & Company
P.O. Box 388
Rochester, VT 05767
1-800-246-8648
www.InnerTraditions.com

Or contact your local bookseller